Light Field Sampling

© Springer Nature Switzerland AG 2022
Reprint of original edition © Morgan & Claypool 2006

Light Field Sampling
Cha Zhang and Tsuhan Chen

ISBN: 978-3-031-01113-9   paperback

ISBN: 978-3-031-02241-8   ebook

DOI 10.1007/978-3-031-02241-8

A Publication in the Springer series
SYNTHESIS LECTURES ON IMAGE, VIDEO, AND MULTIMEDIA PROCESSING #6

Lecture #6
Series Editor: Alan C. Bovik, University of Texas at Austin

Series ISSN: 1559-8136   print
Series ISSN: 1559-8144   electronic

First Edition
10 9 8 7 6 5 4 3 2 1

Light Field Sampling

Cha Zhang
Microsoft Research
Redmond Washington, USA

Tsuhan Chen
Carnegie Melon University
Pittsburgh, Pennsylvania, USA

SYNTHESIS LECTURES ON IMAGE, VIDEO, AND MULTIMEDIA PROCESSING #6

ABSTRACT

Light field is one of the most representative image-based rendering techniques that generate novel virtual views from images instead of 3D models. The light field capture and rendering process can be considered as a procedure of sampling the light rays in the space and interpolating those in novel views. As a result, light field can be studied as a high-dimensional signal sampling problem, which has attracted a lot of research interest and become a convergence point between computer graphics and signal processing, and even computer vision.

This lecture focuses on answering two questions regarding light field sampling, namely how many images are needed for a light field, and if such number is limited, where we should capture them. The book can be divided into three parts.

First, we give a complete analysis on uniform sampling of IBR data. By introducing the surface plenoptic function, we are able to analyze the Fourier spectrum of non-Lambertian and occluded scenes. Given the spectrum, we also apply the generalized sampling theorem on the IBR data, which results in better rendering quality than rectangular sampling for complex scenes. Such uniform sampling analysis provides general guidelines on how the images in IBR should be taken. For instance, it shows that non-Lambertian and occluded scenes often require a higher sampling rate.

Next, we describe a very general sampling framework named freeform sampling. Freeform sampling handles three kinds of problems: sample reduction, minimum sampling rate to meet an error requirement, and minimization of reconstruction error given a fixed number of samples. When the to-be-reconstructed function values are unknown, freeform sampling becomes active sampling. Algorithms of active sampling are developed for light field and show better results than the traditional uniform sampling approach.

Third, we present a self-reconfigurable camera array that we developed, which features a very efficient algorithm for real-time rendering and the ability of automatically reconfiguring the cameras to improve the rendering quality. Both are based on active sampling. Our camera array is able to render dynamic scenes interactively at high quality. To the best of our knowledge, it is the first camera array that can reconfigure the camera positions automatically.

KEYWORDS

Light field, multi-dimensional signal, spectral analysis, sampling, camera array.

Contents

CHAPTER 1

The Light Field

1.1 INTRODUCTION

One might remember that in the movie *Matrix*, the scene with Keanu Reeves dodging the bullets might be one of the most spectacular images ever caught on camera. This filming technology is what the movie producers called *Flo-Mo*. *Flo-Mo* lets the filmmakers shoot scenes where the camera moves at a normal speed while the action is frozen or happens in slow motion. Two movie cameras and 120 computer-controlled still cameras were used in that scene. Similarly, the *Eyevision* system [21] developed by Takeo Kanade [65], which consisted of 33 cameras spaced approximately 6° apart around the rim of the stadium, was used in a live broadcast of Super Bowl game in January 2001. It provided a unique 3D view of selected plays in a 270° stop action image. Unlike traditional 3D rendering techniques which rely on the construction of geometric models to describe the world, these novel viewing experiences were created with tens, hundreds, or even thousands of images. The obvious advantage is that capturing images is often much easier than building complex geometry models for real world scenes. Techniques performing 3D rendering from captured images are widely referred to as *image-based rendering* (IBR), which has been a very active research topic recently.

The idea of image-based rendering can be traced back to the 1930s. A. Gershun defined the phrase *light field* in his classic paper describing the radiometric properties of light in a space [16]. In the early 1990s, Adelson and Bergin proposed the *plenoptic function* (from the Latin root *plenus*, which means complete or full, and *optic*, which pertains to vision), which records the appearance of the whole world, as shown in Fig. 1.1. The plenoptic function is a 7D function that models a 3D dynamic environment by recording the light rays at every space location (V_x, V_y, V_z), toward every possible direction (θ, ϕ), over any range of wavelengths (λ) and at any time (t), i.e.,

$$l(V_x, V_y, V_z, \theta, \phi, \lambda, t). \qquad (1.1)$$

As pointed out by Adelson and Bergen [1]:

> The world is made of three-dimensional objects, but these objects do not communicate their properties directly to an observer. Rather, the objects fill the space around them with

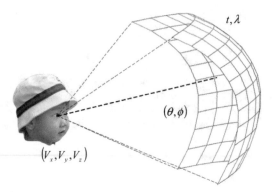

FIGURE 1.1: The 7D plenoptic function

the pattern of light rays that constitutes the plenoptic function, and the observer takes samples from this function. The plenoptic function serves as the sole communication link between the physical objects and their corresponding retinal images. It is the intermediary between the world and the eye.

The light field and the plenoptic function are indeed equivalent [29]. We will use the term *light field* throughout this lecture, and treat the word *plenoptic function* as the mathematical description of the light field.

When we take an image for a scene with a pinhole camera, the light rays passing through the camera's center-of-projection are recorded. They can be considered as samples of the light field/plenoptic function. Image-based rendering can thus be defined under the above plenoptic function framework as follows:

Definition 1.1.1. *Given a continuous plenoptic function that describes a light field,* image-based rendering *is a process of two stages—sampling and rendering. In the sampling stage, samples are taken from the plenoptic function for representation and storage. In the rendering stage, the continuous plenoptic function is reconstructed from the captured samples.*

The above definition reminds us about what we typically do in signal processing: given a continuous signal, sample it and then reconstruct it. The uniqueness of IBR is that the plenoptic function is 7D—a dimension beyond most of the signals handled earlier. In fact, the 7D function is so general that, due to the tremendous amount of data required, no one has been able to sample the full function into one representation. Research on IBR is mostly about how to make reasonable assumptions to reduce the sample data size while keeping reasonable rendering quality.

1.2   THE 4D LIGHT FIELD

There have been many IBR representations invented in the literature. They basically follow two major strategies in order to reduce the data size. First, one may constrain the viewing space of the viewers. Such constraints will effectively reduce the dimension of the plenoptic function, which makes sampling and rendering manageable. For example, if we limit the viewers' interest to static scenes, the time dimension in the plenoptic function can be simply dropped. Second, one may introduce some source descriptions into IBR, such as the scene geometry. Source description has the benefit that it can be very compact. A hybrid source–appearance description is definitely attractive for reducing the data size. To obtain the source description, manual work may be involved or we may resort to computer vision techniques. Interested readers are referred to [23, 80, 59, 44] for general surveys on image-based rendering techniques.

In this lecture, we will focus our attention on the 4D light field, first proposed by Levoy and Hanrahan [29] in 1996. The 4D light field made three reasonable assumptions about the plenoptic function, and is often considered as one of the most classic representations for image-based rendering. These assumptions are:

1. As we are taking images of the scene, we may simplify the wavelength dimension into three channels, i.e., red, green, and blue channels. Each channel represents the integration of the plenoptic function over a certain wavelength range. This simplification can be carried out throughout the capturing and rendering of the scene without noticeable effects.

2. The air is transparent and the radiances along a light ray through empty space remain constant. Under this assumption, we do not need to record the radiances of a light ray on different positions along its path, as they are all identical. To see how we can make use of this assumption, let us limit our interest to the light rays leaving the convex hull of a bounded scene (if the viewer is constrained in a bounded free-space region, the discussion hereafter still applies). Under Assumption 2, the plenoptic function can be represented by its values along an arbitrary surface surrounding the scene. This reduces the dimension of the plenoptic function by 1. The radiance of any light ray in the space can always be obtained by tracing it back to the selected surface. In other words, Assumption 2 allows us to capture a scene at some places and render it somewhere else. It is also one of the most important observations made in [29].

3. The scene is static, thus the time dimension can be dropped. Although a dynamic scene includes much more information than a static one, there are practical concerns that restrict the popularity of dynamic IBR. For instance, we all know that if we capture a video for a scene instead of a single image, the amount of data may increase by about two or three orders of magnitude. It can be expected that dynamic IBR will have the

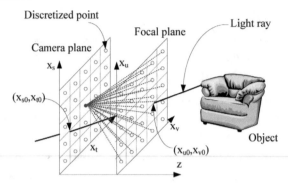

FIGURE 1.2: One parameterization of the 4D light field

same order of size increase from static IBR. Another great benefit of static scene is that one can capture images at different time and positions and use them all to render novel views.

By making these three assumptions, the 7D plenoptic function can be simplified to 4D as follows. Light rays are recorded by their intersections with two planes. One of the planes is indexed with coordinate (x_u, x_v) and the other with coordinate (x_s, x_t), i.e.,

$$l(\mathbf{x}) = l(x_s, x_t, x_u, x_v). \tag{1.2}$$

In Fig. 1.2, we show an example where the two planes, namely the camera plane and the focal plane, are parallel. This is the most typical setup. The two planes are then discretized so that a finite number of light rays are recorded. If we connect all the discretized points from the focal plane to one discretized point on the camera plane, as shown in Fig. 1.2, we get an image (2D array of light rays). Therefore, the 4D representation is also a 2D image array, as is shown in Fig. 1.3.

To create a new view of the object, we first split the view into its light rays, as shown in Fig. 1.4, which are then calculated by interpolating the existing nearby light rays in the image array. For example, let the to-be-rendered light ray in Fig. 1.4 be $(x_{s0}, x_{t0}, x_{u0}, x_{v0})$. Let the nearby discrete points on the camera plane and focal plane be (x_{si}, x_{tj}), $i, j \in \{1, 2\}$, and (x_{uk}, x_{vl}), $k, l \in \{1, 2\}$, respectively. A widely used method to compute the intensity of the to be rendered light ray is through quad-linear interpolation:

$$\hat{l}(x_{s0}, x_{t0}, x_{u0}, x_{v0}) = \sum_{i,j,k,l \in \{1,2\}} \alpha_{i,j,k,l} l(x_{si}, x_{tj}, x_{uk}, x_{vl}), \tag{1.3}$$

where $\alpha_{i,j,k,l}$ are the interpolation weights:

$$\alpha_{i,j,k,l} = (1 - |x_{si} - x_{s0}|)(1 - |x_{tj} - x_{t0}|)(1 - |x_{uk} - x_{u0}|)(1 - |x_{vl} - x_{v0}|), \tag{1.4}$$

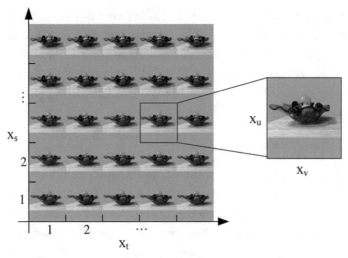

FIGURE 1.3: A sample light field image array: fruit plate

assuming the discrete samples on the two planes are taken on the unit grid. Similar to the well-known linear interpolation method for 1D signals, quad-linear interpolation is not optimal for light field reconstruction, but it is widely adopted for its simplicity. We will discuss more on the reconstruction filter of light field in Chapter 3.

After each light ray is calculated, the new view can be generated simply by reassembling the split rays together. The light field rendering process can be done in real time on modern computers [28, 61]. Its complexity is independent of the scene complexity, and is proportional to the rendered image size. This is an important advantage of light field over traditional model-based rendering techniques, in particular for complex real-world scenes.

As mentioned earlier, when Assumption 2 is made, the plenoptic function can be represented by its values on an arbitrary surface surrounding the scene. Often, that surface is where

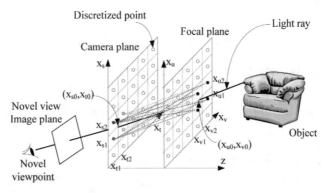

FIGURE 1.4: Rendering of the light field

we put our capturing cameras. The light field in [29] chose this surface to be a box—each face of the box is a camera plane of the two-plane parameterization above. In the spherical light field [20, 7], a spherical surface was chosen for parameterization. Another interesting way to represent all the oriented lines in the space is the sphere–plane light field [7]. In this representation, a light ray is indexed by its direction (2D) and its crossing point (2D) with a plane perpendicular to its direction.

1.3 LIGHT FIELD SAMPLING

In this lecture, our main focus will be on the sampling aspect of the light field. Before going into details, the first question we have to ask is: why do we care about light field sampling?

A regular light field scene requires capturing huge amount of images in order to guarantee perfect reconstruction without aliasing. This is the price one has to pay to eliminate the efforts of building geometry models. While computer memory and hard drive spaces are increasing rapidly, light field is still beyond the capability of most of today's computers. For example, a medium resolution light field may contain 64×64 images, each with 512×512 pixels. The total amount of storage space needed is

$$512 \times 512 \times 64 \times 64 \times 3 = 3 \times 2^{30} \text{ bytes}$$

or 3 gigabytes. Here we assume that each pixel is represented by 3 bytes for three channels. Things can be even worse if people do not know the minimum amount of images needed for a scene, because they tend to capture more images than necessary, causing over-sampling of the scene and wasting even more memory and storage space. The light field sampling theory addresses two problems. First, what is the *minimum* amount of samples/images one has to take for a given scene (or mathematically, a 4D continuous plenoptic function) in order to reconstruct the scene *perfectly*. Second, where to take samples for the scene so that the reconstructed light field has the best rendering quality, assuming the number of samples one can take is limited.

To solve the first problem, we present a high-dimensional spectral and sampling analysis for light field in Chapters 2 and 3. The second problem is addressed by an active sampling framework detailed in Chapters 4 and 5. Below is a road map to the rest of the lecture.

Chapter 2—Light field spectral analysis: In this chapter we describe a general framework for light field spectral analysis based on the surface plenoptic function. Under this framework, we derive the Fourier spectrum of light fields, including those with non-Lambertian and occluded objects.

Chapter 3—Light field uniform sampling: Given the Fourier spectrum of the scene derived in Chapter 2, this chapter focuses on how to sample it efficiently. We show how to sample the light field with rectangular and nonrectangular sampling approaches. We also discuss the sampling problem in both continuous and discrete domain. Finally, light field sampling is discussed in the joint image and geometry space, which leads to many potential applications.

Chapter 4—The freeform sampling framework: This chapter tries to give a very general sampling framework named freeform sampling. Samples are no longer limited to regular patterns, thus solutions to freeform sampling also change. Active sampling is a special case of freeform sampling, where the sample values on the reconstruction set are unknown. General algorithms are also presented for active sampling.

Chapter 5—Light field active sampling: We apply the active sampling theorem in the last chapter to light field, including active incremental sampling and active rearranged sampling. Both algorithms demonstrate much improved performance over traditional sampling methods.

Chapter 6—The self-reconfigurable camera array: This chapter presents a real-world system that captures videos from an array of mobile cameras, renders novel views on the fly, and reconfigures the camera positions to achieve better rendering quality. We hope this chapter can bring the readers some insights on how the proposed active sampling schemes can be applied in practice.

CHAPTER 2

Light Field Spectral Analysis

As illustrated in Chapter 1, light field is a four-dimensional signal recording the light rays in the space. The goal of light field spectral analysis is to compute the Fourier transform of the light field signal, so that it can be used to determine the sampling rate of the scene (sampling is discussed in Chapter 3). Denote $l(\mathbf{x})$ as the light field signal, where $\mathbf{x} = [x_s \ x_t \ x_u \ x_v]^T$. The Fourier transform of the light field signal is formally defined as

$$L(\Omega) = \int_{-\infty}^{\infty} l(\mathbf{x})e^{-j\Omega^T \mathbf{x}} d\mathbf{x}, \tag{2.1}$$

where $\Omega = [\Omega_s \ \Omega_t \ \Omega_u \ \Omega_v]^T$. To convert the Fourier transform back to the light field, we have the inverse transform as

$$l(\mathbf{x}) = \frac{1}{(2\pi)^D} \int_{-\infty}^{\infty} L(\Omega)e^{j\Omega^T \mathbf{x}} d\Omega, \tag{2.2}$$

where D is the dimension of the signal, which is 4 in the case of a 4D light field.

The Fourier transform of the light field is affected by a number of factors, including the scene geometry, the texture on the scene surface, the reflection property of the scene surface, the capturing and rendering cameras' resolution, etc. There have been works that attempted to address this problem such as [30, 8, 34]. In this chapter, we present an analytical framework based on the concept of *surface plenoptic function* (SPF), which represents the light rays starting from the object surface. The new framework is more generic than previous approaches, hence we can derive the sampling rate of the light the field for scenes with Lambertian or non-Lambertian surfaces and with or without occlusions.

2.1 THE SURFACE PLENOPTIC FUNCTION

We first introduce the surface plenoptic function, which is another 4D representation of the light field under the same assumptions made in Section 1.2. Any light ray in the free space has a source. It can be either emitted from some light source (e.g., the Sun) or reflected by some object surface. If the object is transparent, refraction is also involved. Let the entire surface of all the light sources and objects be S. We can always trace a light ray in the free space back to

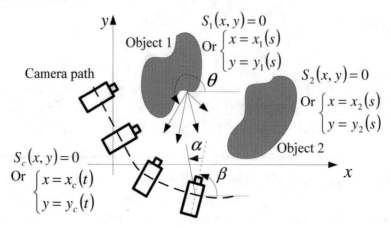

FIGURE 2.1: The 2D surface plenoptic function and general IBR capturing

a point on S. The 7D plenoptic function can thus be re-parameterized to 4D including point on the surface S (2D) and azimuth and elevation angles (2D) the light ray is emitted, assuming the assumptions made in Section 1.2 hold. We name this simplified function as the 4D *surface plenoptic function* (SPF) [37, 74].

   As an independent representation of the light field, Wood et al. [74] have discussed the representation, rendering, and compression of surface light field. Wong's plenoptic illumination function [73] and Lin's reflected irradiance field [31] are also similar to SPF in the sense that they tried to model different light ray radiances from same surface points. On the other hand, SPF is used in this book for the sampling analysis of the standard 4D light field. The basic observation is, since light rays start from the object surface and end at the capturing cameras, there exists an onto mapping from the SPF to the standard light field. Such a mapping depends on both the scene geometry and the camera surface, but not the surface property such as the bidirectional reflection distribution function (BRDF). If we have some knowledge about the scene, in other words, if we know some property about the SPF, related property can be derived for the standard 4D light field.

   Let us first give a brief look at the mapping between SPF and general IBR representations. Without loss of generality, we use the 2D world as an example throughout this chapter for conciseness. The conclusions drawn here are easy to extend to the 3D world. In the 2D world, surface of objects/light sources is described with curves. Ignoring time and wavelength, the SPF is 2D: one dimension for describing a point on a curve, the other for illustrating the direction of the light ray emitted/reflected/refracted. An example scene is shown in Fig. 2.1. The surface can be represented by

$$S_i(x, y) = 0 \quad \text{or} \quad \begin{cases} x = x_i(s) \\ y = y_i(s), \end{cases} \qquad (2.3)$$

where s is the arc length and i is the index for different objects. For a certain object i, we define its SPF as

$$l_i(s, \theta) \text{ on the curve} \begin{cases} x = x_i(s) \\ y = y_i(s), \end{cases} \tag{2.4}$$

where $0 \leq \theta < 2\pi$ is the direction of the light ray. $l_i(s, \theta)$ is the radiance of the light ray that can be traced back to the surface point determined by s on object i. Note that the above function does not appear to be related with what people often use for calculating lightings, such as surface normal, BRDF, etc. We intend to do so because it is often too complicated if we try to model how the light rays are generated, in addition to the fact that such a model does not always exist for real scenes. Therefore, we only consider the resultant lighting effects in (2.4), and assume that we have some knowledge about the SPF. If we are able to model the lighting very well, the following analysis can still apply after calculating the lighting effects based on the known model.

Lambertian is the first assumption we could make for the scene, since it has been exclusively used in the IBR sampling literature. In terms of SPF, Lambertian property gives the following relationship:

$$l_i(s, \theta) = l_{is}(s); \tag{2.5}$$

and its Fourier transform is

$$L_i(\Omega_s, \Omega_\theta) = L_{is}(\Omega_s)\delta(\Omega_\theta), \tag{2.6}$$

where $L_i(\Omega_s, \Omega_\theta)$ is the Fourier transform of $l_i(s, \theta)$ and $L_{is}(\Omega_s)$ is the Fourier transform of $l_{is}(s)$.[1]

In the real world, pure Lambertian objects are rare. Highly reflective surface (like a mirror) is infrequent, too. In most cases, light rays from the same point on the object surface tend to be similar and have slow changes with respect to their angles. It is therefore reasonable to assume that $L_i(\Omega_s, \Omega_\theta)$ can be approximated by a band-limited signal. That is,

$$L_i(\Omega_s, \Omega_\theta) \approx L_i(\Omega_s, \Omega_\theta)I_{B_i}(\Omega_\theta), \tag{2.7}$$

where $I_{B_i}(\Omega_\theta)$ is the indicator function over Ω_θ, which is defined as

$$I_{B_i}(\Omega_\theta) = \begin{cases} 1, & \text{if } |\Omega_\theta| < B_i; \\ 0, & \text{otherwise.} \end{cases} \tag{2.8}$$

---

[1]Strictly speaking, at a certain point on the surface, $l_i(s, \theta)$ need to be truncated on θ because we can only observe light rays that come out of the object. Therefore, along Ω_θ $L_i(\Omega_s, \Omega_\theta)$ cannot be a delta function, nor can it be band limited. However, we assume that this windowing artifact is negligible. We do consider artifacts caused by self or mutual occlusions, as presented in Section 2.2.3.

Here B_i defines the bandwidth for object i. This band-limitedness assumption can be connected to the band limitedness of the surface BRDF with the signal-processing framework for inverse rendering presented by Ramamoorthi and Hanrahan [51]. For points on a reflective surface, the outgoing light can be described as the convolution of the incoming light and the surface BRDF. If the incoming light is far (thus the incoming light can be considered as a delta function with respect to the angle), as long as the BRDF is band limited, the outgoing light rays will be band limited.

In order to capture the plenoptic function or surface plenoptic function, the existing IBR approaches align cameras on a path/surface and take images for the scene. For example, cameras are placed on a plane in light field [29], and on a circle in concentric mosaics [58]. In the 2D world, 2D light field has cameras on a line, while 2D concentric mosaics have cameras on a circle. In general, the cameras can be put along an arbitrary curve, as is shown in Fig. 2.1. Let the camera path be

$$S_c(x, y) = 0 \quad \text{or} \quad \begin{cases} x = x_c(t) \\ y = y_c(t), \end{cases} \tag{2.9}$$

where t is the arc length. Assume that the camera path curve is differentiable, and the optical axes of our cameras are identical to the normals of the camera path. That is, at arc length t, the optical axis has the direction $(-y_c'(t), x_c'(t))$, where $x_c'(t)$ and $y_c'(t)$ are the first-order derivatives. Denote the direction of the optical axis with angle β; then

$$\tan(\beta) = -\frac{x_c'(t)}{y_c'(t)}. \tag{2.10}$$

The image pixels can be indexed by the angle between the captured light ray and the optical axis, as is represented by α in Fig. 2.1. Denote the radiance of the light ray captured at arc length t, angle α as $l_c(t, \alpha)$. The goal of spectral analysis for this specific IBR representation is to find the Fourier transform of $l_c(t, \alpha)$, denoted as $L_c(\Omega_t, \Omega_\alpha)$, so that we can determine the minimum number of images we have to capture. Given the knowledge we have on the SPF $l_i(s, \theta)$ and its Fourier transform $L_i(\Omega_s, \Omega_\theta)$, the strategy is to associate $l_c(t, \alpha)$ with $l_i(s, \theta)$ and hope that we can represent $L_c(\Omega_t, \Omega_\alpha)$ in terms of $L_i(\Omega_s, \Omega_\theta)$.

2.2   ANALYSIS OF SCENES WITH KNOWN GEOMETRY

As shown in Fig. 2.2, the 2D light field is parameterized by two parallel lines, indexed by x_t and x_v, respectively. The x_t line is the camera line, while the x_v line is the focal line. The distance between the two lines is f, which is the focal length of the cameras. It is easy to show that a light ray indexed by pair (x_t, x_v) satisfies the following algebraic equation:

$$fx - x_v y - fx_t = 0. \tag{2.11}$$

Note that the focal line is indexed locally with respect to where the camera is.

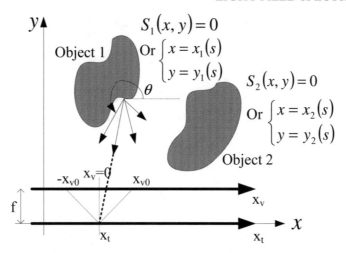

FIGURE 2.2: The light field parameterization with surface plenoptic function

The relationship between the light field $l(x_t, x_v)$ and the SPF $l_i(s, \theta)$ is as follows. For the same light ray emitted/reflected/refracted from a surface point, it must be captured at the corresponding angle. That is,

$$\tan(\theta) = \frac{f}{x_v} \quad \text{or} \quad \theta = \frac{3\pi}{2} - \tan^{-1}\left(\frac{x_v}{f}\right), \tag{2.12}$$

where $-x_{v0} \le x_v \le x_{v0}$ and $2\tan^{-1}\left(\frac{x_{v0}}{f}\right)$ is the field of view (FOV). The above equation is actually the mapping between a pixel's angular position θ and its image coordinate x_v. Such a mapping can be linearized as

$$\theta \approx \frac{3\pi}{2} - \frac{x_v}{f}. \tag{2.13}$$

The above linearization will introduce a 4.3% maximum error if the FOV of the camera is 40°. In practice, the approximation in Eq. (2.13) can be replaced by a simple pixel rearrangement. In the following discussions, we denote $\phi = \frac{x_v}{f} \approx \frac{3\pi}{2} - \theta$ for conciseness, where $-\frac{x_{v0}}{f} \le \phi \le \frac{x_{v0}}{f}$ due to the limited FOV.

Another constraint is that the light ray (x_t, x_v) can be traced back to a cross point on the object surface, whose arc length s can be obtained through solving

$$\begin{cases} x = x_i(s) \\ y = y_i(s) \\ fx - x_v y - fx_t = 0. \end{cases} \tag{2.14}$$

When multiple objects exist in the scene or some objects occlude themselves, Eq. (2.14) may have multiple answers. We have to figure out which cross point is the closest to the cameras. The closest point will occlude all the others. This may make scenes with occlusions hard to analyze. However, for simple scenes this is still doable and we will show examples later in this chapter.

2.2.1   Scene at a Constant Depth

The simplest scene we can have for the light field is one at a constant depth, as shown in Fig. 2.3(a). Denote its SPF as $l_0(s, \theta)$. The surface can be described by

$$\begin{cases} x = x_0(s) = s \\ y = y_0(s) = d_0. \end{cases} \tag{2.15}$$

We can solve Eq. (2.14) without concerning about occlusion:

$$fs - vd_0 - ft = 0 \quad \Longrightarrow \quad s = \frac{vd_0}{f} + t. \tag{2.16}$$

The light field spectrum can be derived as

$$\begin{aligned} L(\Omega_t, \Omega_v) &= \iint l(x_t, x_v)e^{-j\Omega_t x_t - j\Omega_v x_v}\mathrm{d}x_t \mathrm{d}x_v \\ &\approx fL_0(\Omega_t, d_0\Omega_t - f\Omega_v)e^{j\frac{3\pi}{2}(d_0\Omega_t - f\Omega_v)}, \end{aligned} \tag{2.17}$$

where the approximation is due to Eq. (2.13). We can see that the spectrum of the light field at a constant depth is a rotated version of the SPF spectrum, with some constant factor in magnitude and some shift in phase. The rotation slope is determined by the scene depth d_0 and the focal length f as $\frac{\Omega_v}{\Omega_t} = \frac{d_0}{f}$.

If the object surface is Lambertian, we have $L_0(\Omega_s, \Omega_\theta) = L_{0s}(\Omega_s)\delta(\Omega_\theta)$ as stated in Eq. (2.6). Therefore,

$$\begin{aligned} L(\Omega_t, \Omega_v) &= fL_0(\Omega_t, d_0\Omega_t - f\Omega_v)e^{j\frac{3\pi}{2}(d_0\Omega_t - f\Omega_v)} \\ &= fL_{0s}(\Omega_t)\delta(d_0\Omega_t - f\Omega_v)e^{j\frac{3\pi}{2}(d_0\Omega_t - f\Omega_v)}, \end{aligned} \tag{2.18}$$

which is clearly a tilted line in the (Ω_t, Ω_v) space.

When the object surface is non-Lambertian but the SPF is band limited, we have $L_0(\Omega_s, \Omega_\theta) \approx L_0(\Omega_s, \Omega_\theta)I_{B_0}(\Omega_\theta)$ as in Eq. (2.7). Consequently,

$$\begin{aligned} L(\Omega_t, \Omega_v) &= fL_0(\Omega_t, d_0\Omega_t - f\Omega_v)e^{j\frac{3\pi}{2}(d_0\Omega_t - f\Omega_v)} \\ &\approx fL_0(\Omega_t, d_0\Omega_t - f\Omega_v)e^{j\frac{3\pi}{2}(d_0\Omega_t - f\Omega_v)}I_{B_0}(d_0\Omega_t - f\Omega_v). \end{aligned} \tag{2.19}$$

The spectrum is also tilted, but this time it has a finite width $\frac{2B_0}{\sqrt{d_0^2 + f^2}}$ perpendicular to the tilted spectrum (or $\frac{2B_0}{d_0}$ horizontally) because of the indicator function.

The above analysis is illustrated in Fig. 2.3. A scene at a constant depth has two sinusoids (different frequency) pasted on it as texture, as shown in Fig. 2.3(a). Fig. 2.3(b) is the epipolar image (EPI, which is an image composed of light rays using the camera line t and the focal line v

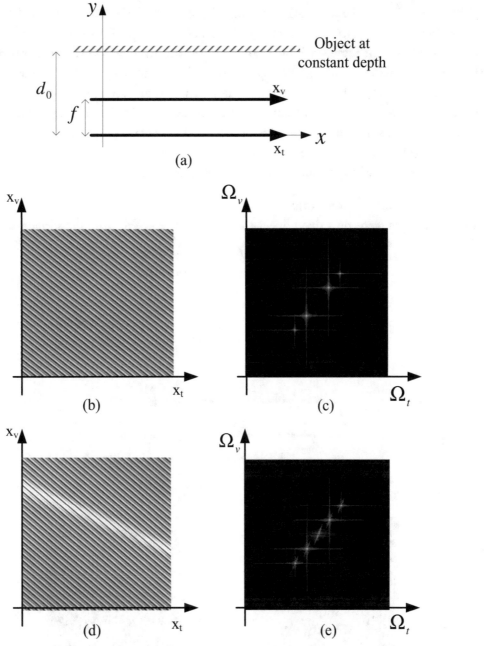

FIGURE 2.3: Spectrum of Lambertian and non-Lambertian scenes at constant depth. (a) A scene at constant depth; (b) the EPI of the light field when the scene is Lambertian; (c) the Fourier transform of (b); (d) the EPI of the light field when the scene is non-Lambertian; (e) the Fourier transform of (d)

as its two axes) when the scene is Lambertian. Fig. 2.3(c) is its Fourier transform. The spectrum has several peaks because the texture on the scene object is purely sinusoidal. It basically lies on a tilted line with some small horizontal and vertical windowing artifact that is due to the truncation of the range of s and θ. We ignored the windowing artifacts in our analysis for simplicity. Fig. 2.3(d) shows the EPI for a non-Lambertian case at the same depth, which is generated by adding a point light source near the scene and assuming a Phong surface reflection model [48]. It can be seen that because of the non-Lambertian property, its Fourier transform in Fig. 2.3(e) is expanded. The direction of the expansion is determined by the depth of the point light source, which will also affect the SPF bandwidth.

2.2.2   Scene on a Titled Line

We next study a scene on a tilted line, as shown in Fig. 2.4. We write the surface equation as

$$\begin{cases} x = x_0(s) = s \cos \varphi + x_0 \\ y = y_0(s) = s \sin \varphi + y_0, \end{cases} \qquad (2.20)$$

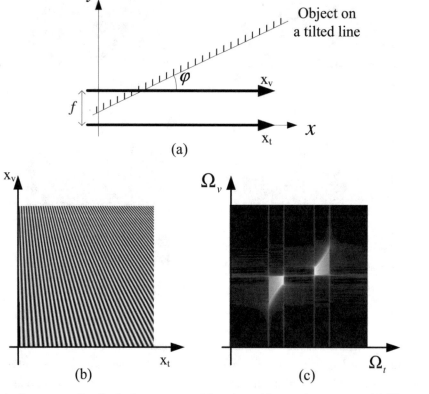

(a)

(b)

(c)

FIGURE 2.4: Spectrum of a tilted planar scene with a sinusoid pasted as texture. (a) The scene on a tilted line; (b) the EPI of the light field; (c) the Fourier transform of (b)

where $0 \leq \varphi < \pi$ is an angle that is known. Assuming no occlusion, we can solve Eq. (2.14) as

$$f(s\cos\varphi + x_0) - x_v(s\sin\varphi + y_0) - fx_t = 0 \implies s = \frac{fx_t - fx_0 + x_v y_0}{f\cos\varphi - x_v\sin\varphi}. \quad (2.21)$$

To make sure that each captured image contains no empty space, φ cannot be arbitrary. If $-x_{v0} \leq x_v \leq x_{v0}$, we need to have a constraint that $|f\operatorname{ctg}\varphi| > x_{v0}$.

It can be shown that the light field spectrum is

$$L(\Omega_t, \Omega_v) = \iint l(x_t, x_v)e^{-j\Omega_t x_t - j\Omega_v x_v}dx_t dx_v$$

$$\approx \iint l_0(s, \theta)e^{-j\Omega_t\left[s(\cos\varphi - \phi\sin\varphi) + (x_0 - \phi y_0)\right] - j\Omega_v\phi f}$$

$$\cdot (\cos\varphi - \phi\sin\varphi)f ds\, d\theta. \quad (2.22)$$

Unfortunately, even for a scene as simple as a line, Eq. (2.22) is too complex to solve. Here we consider a case where Eq. (2.22) can be further simplified. Let the scene be Lambertian, i.e., as in Eq. (2.6). We have,

$$L(\Omega_t, \Omega_v) \approx \iint l_0(s, \theta)e^{-j\Omega_t\left[s(\cos\varphi - \phi\sin\varphi) + (x_0 - \phi y_0)\right] - j\Omega_v\phi f}$$

$$\cdot (\cos\varphi - \phi\sin\varphi)f ds\, d\theta.$$

$$= \int L_{0s}\left[\Omega_t(\cos\varphi - \phi\sin\varphi)\right]e^{-j\Omega_t(x_0 - \phi y_0) - j\Omega_v\phi f}$$

$$\cdot (\cos\varphi - \phi\sin\varphi)f d\theta. \quad (2.23)$$

If we have a sinusoid pasted as the texture, e.g., $l_{0s}(s) = \sin(\Omega_0 s)$, then its Fourier transform is

$$L_{0s}(\Omega_s) = \frac{1}{2j}\left[\delta(\Omega_s - \Omega_0) - \delta(\Omega_s + \Omega_0)\right]. \quad (2.24)$$

Due to symmetry, let us consider the magnitude of $L_c(\Omega_t, \Omega_v)$ when $\Omega_t > 0$ and $\Omega_s > 0$. From Eq. (2.23), we have,

$$\left|L(\Omega_t, \Omega_v)\right| = \begin{cases} \frac{\Omega_0 f}{2\Omega_t^2|\sin\varphi|}, & \text{when } \frac{\Omega_0 f}{f\cos\varphi + v_0|\sin\varphi|} \leq \Omega_t \leq \frac{\Omega_0 f}{f\cos\varphi - v_0|\sin\varphi|}; \\ 0, & \text{otherwise.} \end{cases}$$

$$(2.25)$$

Fig. 2.4(a) shows a scene on a line with $\varphi = \frac{\pi}{6}$. A single sinusoid is pasted on the line as texture. Fig. 2.4(b) is the EPI of the light field, and Fig. 2.4(c) is its Fourier transform. Note that the theoretical analysis matches very well with the example. For instance, the spectrum is nonzero only between certain Ω_t thresholds. In the nonzero region, the spectrum decays when $|\Omega_t|$ increases.

2.2.3   Occlusions Between Objects

When occlusion happens, light field spectral analysis becomes very difficult due to the discontinuity at the occlusion boundary. In fact, theoretically any occlusion will cause the Fourier spectrum to be band unlimited. In this subsection, we analyze occluded scenes under certain assumptions, and give some insights on the formation of the spectrum of such scenes.

With SPF, solving for occluded scenes means finding the closest cross point among multiple solutions for Eq. (2.14). We first assume that the objects do not occlude themselves. Marchand-Maillet [34] once derived the no-occlusion condition for functional surfaces. Similarly, in our notation, no occlusion requires

$$\max_s \left| \frac{y_i'(s)}{x_i'(s)} \right| < \frac{f}{v_0}, \tag{2.26}$$

where $y_i'(s)$ and $x_i'(s)$ are first-order derivatives. Equation (2.26) simply means that the slope of the surface should not be greater than that of any possible light rays captured. This assumption is, however, hard to justify. We treat this as an example where occlusion can be solved under our parameterization. In the meantime, in practice mutual occlusions are often more significant than self-occlusions[2], and self-occluded objects can sometimes be decomposed into smaller objects so that the occlusions become mutual.

When objects do not occlude themselves, their spectrums can be obtained through previously mentioned methods. Let the number of objects in the scene be N. Let $l_i(x_t, x_v)$, $0 \le i \le N - 1$, be the N light fields and $L_i(\Omega_t, \Omega_v), 0 \le i \le N - 1$, be their Fourier transforms. We also define a silhouette light field for each object i as

$$\gamma_i(x_t, x_v) = \begin{cases} 1, & \text{when light ray } (x_t, x_v) \text{ can be traced back to object } i; \\ 0, & \text{otherwise.} \end{cases} \tag{2.27}$$

Their Fourier transforms are denoted as $\Gamma_i(\Omega_t, \Omega_v), 0 \le i \le N - 1$. Note that the silhouette's spectrum can be obtained by setting $l_i(s, \theta) \equiv 1$ on object i's surface.

We are now ready to find the spectrum of the occluded scene. Since Fourier transform is linear, the overall spectrum is simply the sum of individual objects' spectrums. If an object is not occluded, we can just add its spectrum to the overall one. Otherwise, let object i be occluded by K other objects ($K < N - 1$). Denote the K occluding objects' silhouette light fields as $\gamma_{i_k}(x_t, x_v), 0 \le k \le K - 1$. The contribution object i has to the overall light field can

---

[2]We claim mutual occlusion is more significant than self-occlusion because the former often causes a sharp boundary/ edge in the EPI, while self-occlusion often does not if the surface normal changes slowly.

be written as

$$\tilde{l}_i(x_t, x_v) = l_i(x_t, x_v) \prod_{k=0}^{K-1} \left[1 - \gamma_{i_k}(x_t, x_v)\right], \qquad (2.28)$$

where $\tilde{l}_i(t, v)$ is the occluded light field of object i. Its Fourier transform is

$$\widetilde{L}_i(\Omega_t, \Omega_v) = L_i(\Omega_t, \Omega_v) \otimes \left[\delta(\Omega_t, \Omega_v) - \Gamma_{i_0}(\Omega_t, \Omega_v)\right]$$
$$\otimes \cdots \otimes \left[\delta(\Omega_t, \Omega_v) - \Gamma_{i_{K-1}}(\Omega_t, \Omega_v)\right], \qquad (2.29)$$

where $\otimes$ stands for convolution, $\delta(\Omega_t, \Omega_v)$ is the Fourier transform of constant 1. From Eq. (2.29) we can see that the spectrum of an occluded object is its unoccluded spectrum modulated by all the occluding objects' silhouette spectrum. This modulation will bring some additional components to the overall spectrum.

Fig. 2.5(a) shows an example Lambertian scene that has three objects. Each object is at a constant depth and has two sinusoids pasted as texture. Therefore, from what we had

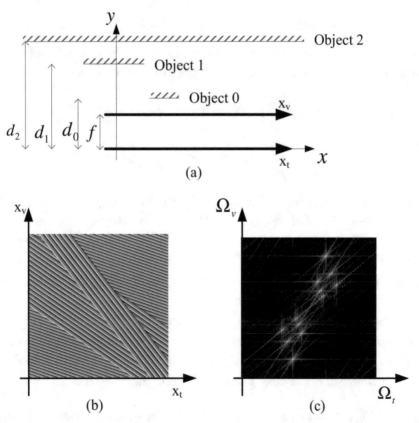

(a)

(b)

(c)

FIGURE 2.5: Spectrum of an occluded scene. (a) Three objects on different constant depths; (b) the EPI of the light field; (c) the Fourier transform of (b)

in Section 2.2.1, if no occlusion is considered, for $0 \leq i \leq 2$, $L_i(\Omega_t, \Omega_v)$ and $\Gamma_i(\Omega_t, \Omega_v)$ both lie on a tilted line whose slope is $\frac{d_i}{f}$, where d_i is the depth of object i. Since objects at larger depth are occluded by closer objects, we will notice additional modulated components along larger slope lines. Moreover, the additional modulated components will be along smaller slopes corresponding to the closer object. This is clearly shown in Fig. 2.5(c).

2.3   ANALYSIS OF SCENES WITH UNKNOWN GEOMETRY

Practically, it is often hard to measure the geometry of the scene objects in order to derive the spectrum of the light field. After all, one of the major advantages of image-based rendering over model-based rendering is the capability to render 3D views without knowing the scene geometry. In this section we discuss the spectral analysis for scenes with unknown geometry.

A complicated environment can be approximated using truncated and piecewise constant depth segments $d_j, 0 \leq j \leq J - 1$. For each segment, if it is not occluded by other segments, the analysis in Section 2.2.1 applies, and its Fourier transform can be written as

$$L_j(\Omega_t, \Omega_v) \approx f L_{0j}(\Omega_t, d_j\Omega_t - f\Omega_v)e^{j\frac{3\pi}{2}(d_j\Omega_t - f\Omega_v)}, \qquad (2.30)$$

where $L_{0j}(\Omega_s, \Omega_\theta)$ is the Fourier transform of the SPF on that segment. Due to the additivity of the Fourier transform, the whole scene's spectrum can be written as

$$
\begin{aligned}
L(\Omega_t, \Omega_v) &= \sum_{j=0}^{J-1} L_j(\Omega_t, \Omega_v) \\
&\approx \sum_{j=0}^{J-1} f L_{0j}(\Omega_t, d_j\Omega_t - f\Omega_v)e^{j\frac{3\pi}{2}(d_j\Omega_t - f\Omega_v)}.
\end{aligned}
\qquad (2.31)
$$

Let the scene be Lambertian. Following Eq. (2.18), $L_j(\Omega_t, \Omega_v)$ is a tilted line in the (Ω_t, Ω_v) space[3]. Hence the whole spectrum will be between two slopes, one determined by the minimum scene depth d_{min} and the other determined by the maximum scene depth d_{max}. This is illustrated in Fig. 2.6(a). Note that in Fig. 2.6(a), we let the unit of the x_v axis be 1, so that $-\pi \leq \Omega_v < \pi$. With today's high resolution cameras, the sampling rate of the x_v axis is often much higher than the sampling rate of the x_t axis.

If the scene is non-Lambertian, it can be expected that for each segment of depth d_j, the spectrum support will be a "fatter" ellipse. This causes the original fan-like spectrum to be expanded. If the SPF is band limited along Ω_θ as in Eq. (2.7), the light field bandwidth expansion is also limited, as shown in Fig. 2.6(b). The dotted region stands for the extra spectrum support

---

[3] Strictly speaking, $L_j(\Omega_t, \Omega_v)$ is not a line but rather an ellipse, because spatially the object at depth d_j is truncated. In the real world this will not affect too much on the final spectrum support, as demonstrated in [8].

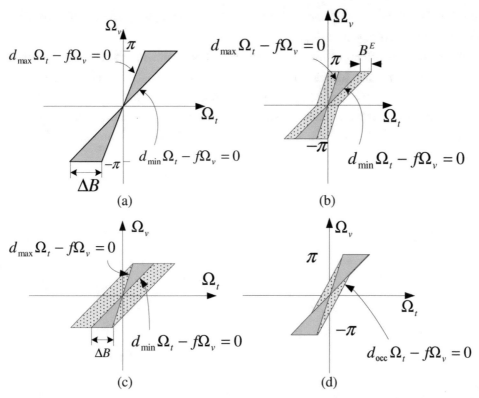

FIGURE 2.6: Spectrum of scenes with unknown geometry. (a) The spectrum of a nonoccluded and Lambertian scene. (b) The spectrum of a non-Lambertian scene. (c) The most conservative estimation of the spectrum of an occluded scene. (d) The spectrum of an occluded scene given that the object at d_{occ} is the main source of occlusion

caused by the non-Lambertian surface. A *conservative* estimation of the amount of expansion is

$$B^E = \max_j \frac{B_j}{d_j} \qquad (2.32)$$

on each side of the spectrum along the horizontal axis Ω_t, based on the discussion after Eq. (2.19). Here B_j and d_j represent the bandwidth of SPF and depth for object depth layer j.

The above analysis can be applied to more complex scenes if occlusions have to be considered. When the scene is heavily occluded, far layers will be blocked by close layers, which causes a spectrum modulation as in Eq. (2.29). In worst cases, the major occluding objects are at d_{min}, which generates additional modulated components along the slope $\frac{d_{\text{min}}}{f}$, where f is the focal length. The corresponding spectrum is given in Fig. 2.6(c). The dotted support is the additional modulated components. Fig. 2.6(d) illustrates the spectrum of a scene which is mainly occluded by an object at d_{occ}.

In practice, the effects of non-Lambertian and occlusions in expanding the spectrum is often negligible for regular scenes. In other words, it is often safe to claim that the spectrum of the light field is simply bounded by the minimum and maximum scene depth. On the other hand, for scenes that contains reflective surface or very heavy occlusions, the expansion of the spectrum is significant. We will show some examples in the next chapter when discussing the sampling procedure.

CHAPTER 3

Light Field Uniform Sampling

The spectrum analysis discussed in the previous chapter lays the foundation of light field sampling. The next step is to sample the light field according to the spectrum and make sure there is no aliasing during this process, so that we can reconstruct the continuous light field with no artifacts. Early works on light field sampling mostly considered rectangular sampling [30, 8]. As a well-known fact in multidimensional signal processing theory [11], rectangular sampling is not the best sampling strategy in high-dimensional space, where the light field stands. In this chapter, we study the sampling process of the light field with the generalized sampling theory and treat rectangular sampling as a special case, which turns out to be the best choice for real-world scenes in terms of the tradeoff between complexity and accuracy.

3.1 THE SAMPLING THEORY FOR MULTIDIMENSIONAL SIGNALS

3.1.1 Continuous Domain Sampling

For a multidimensional signal such as light field, periodic samples can be taken with arbitrary sampling geometry, as shown in Fig. 3.1. Take the 2D case as an example. Let $l(\mathbf{x})$ be the spatial domain continuous signal and $L(\Omega)$ be its Fourier transform. Define the sampling matrix $\mathbf{V}$ as

$$\mathbf{V} = [\mathbf{v}_1 \, \mathbf{v}_2], \qquad (3.1)$$

where $\mathbf{v}_1$ and $\mathbf{v}_2$ are two linearly independent vectors. Note that the traditional rectangular sampling strategy is a special case here, which corresponds to the sampling matrices that are diagonal.

Sampling the continuous signal $l(\mathbf{x})$ with matrix $\mathbf{V}$ produces the discrete signal

$$l_d(\mathbf{n}) = l(\mathbf{V}\mathbf{n}), \qquad (3.2)$$

where $l_d(\mathbf{n})$ stands for the sampled signal and $\mathbf{n} = [n_1 \, n_2]^T \in \mathcal{N}$ is an integer vector representing the index of the samples along $\mathbf{v}_1$ and $\mathbf{v}_2$. The inverse of the determinant of $\mathbf{V}$, namely $\frac{1}{|\mathbf{V}|}$, represents the density of samples per unit area, which should be as small as possible to save the number of samples as long as there is no aliasing.

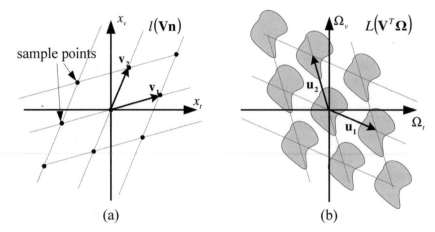

FIGURE 3.1: Generalized sampling theorem in 2D space. (a) Spatial sampling on a lattice with arbitrary geometry. (b) Frequency domain replication along a corresponding lattice

Given the sampled light field, we can define its discrete-time Fourier transform as

$$L_d(\omega) = \sum_{\mathbf{n} \in \mathcal{N}} l_d(\mathbf{n}) e^{-j\omega^T \mathbf{n}} \tag{3.3}$$

if the summation converges. The inverse transform is given by

$$l_d(\mathbf{n}) = \frac{1}{(2\pi)^D} \int_{\omega \in [-\pi, \pi)^D} L_d(\omega) e^{j\omega^T \mathbf{n}} d\omega, \tag{3.4}$$

where $[-\pi, \pi)^D$ denotes the set of $D \times 1$ real vectors ω with components ω_i in the range $-\pi \leq \omega_i < \pi$.

There is a close relationship between the continuous Fourier transform $L(\Omega)$ and the discrete-time Fourier transform of the sampled signal $L_d(\omega)$ [11],

$$L_d(\omega) = \frac{1}{|\mathbf{V}|} \sum_{\mathbf{k} \in \mathcal{N}} L\big(\mathbf{V}^{-T}(\omega - 2\pi \mathbf{k})\big)$$

$$\text{or} \quad L_d(\mathbf{V}^T \Omega) = \frac{1}{|\mathbf{V}|} \sum_{\mathbf{k} \in \mathcal{N}} L(\Omega - \mathbf{U}\mathbf{k}), \tag{3.5}$$

where $\mathbf{k} = [k_1 \, k_2]^T \in \mathcal{N}$; $\mathbf{U}$ is a matrix that satisfies

$$\mathbf{U}^T \mathbf{V} = 2\pi \mathbf{I}, \tag{3.6}$$

and $\mathbf{I}$ is the identity matrix. From Eq. (3.5), we can see that periodic sampling on a parallelepiped lattice in the spatial domain leads to the replication of the Fourier transform on another parallelepiped lattice in the frequency domain. This is clearly shown in Fig. 3.1, where the two lattices satisfy Eq. (3.6). Note that if we perform rectangular sampling, both $\mathbf{V}$ and $\mathbf{U}$ are diagonal matrices, and the discrete-time spectrum will be duplicated on a grid that is parallel to the main axes.

3.1.2   Discrete Domain Sampling

Sometimes what we begin with is already a discrete signal. It is possible to perform sampling on discrete signals, too. For example, the $\mathbf{M}$-fold down-sampled version of $l_d(\mathbf{n})$ is defined as $r_d(\mathbf{n}) = l_d(\mathbf{Mn})$, where $\mathbf{M}$ is a nonsingular integer matrix called the down-sampling matrix. In the frequency domain, the relationship is [11]

$$R_d(\omega) = \frac{1}{|\mathbf{M}|} \sum_{\mathbf{k} \in \mathcal{N}(\mathbf{M}^T)} L_d\left(\mathbf{M}^{-T}(\omega - 2\pi\mathbf{k})\right)$$

$$\text{or} \quad R_d(\mathbf{M}^T\omega) = \frac{1}{|\mathbf{M}|} \sum_{\mathbf{k} \in \mathcal{N}(\mathbf{M}^T)} L_d(\omega - \mathbf{Nk}), \qquad (3.7)$$

where $\mathcal{N}(\mathbf{M}^T)$ is the set of all integer vectors of the form $\mathbf{M}^T\mathbf{x}, \mathbf{x} \in [0, 1)^D$. Similar to the continuous case, in the frequency domain, down-sampling the signal with $\mathbf{M}$ is equivalent to duplicating the original spectrum on the grid $\mathbf{Nk}$ with $\mathbf{N} = 2\pi\mathbf{M}^{-T}$ and $\mathbf{k} \in \mathcal{N}(\mathbf{M}^T)$, followed by transforming the whole spectrum with matrix $\mathbf{M}^{-T}$.

Discrete domain down-sampling shall normally be performed after prefiltering to avoid aliasing. This is shown in Fig. 3.2(a). Given the sampling matrix $\mathbf{M}$, the filter typically have parallelepiped-shaped passband support in the region

$$\omega = \pi\mathbf{M}^{-T}\mathbf{x} + 2\pi\mathbf{k}, \qquad \mathbf{x} \in [-1, 1)^D, \quad \mathbf{k} \in \mathcal{N}, \qquad (3.8)$$

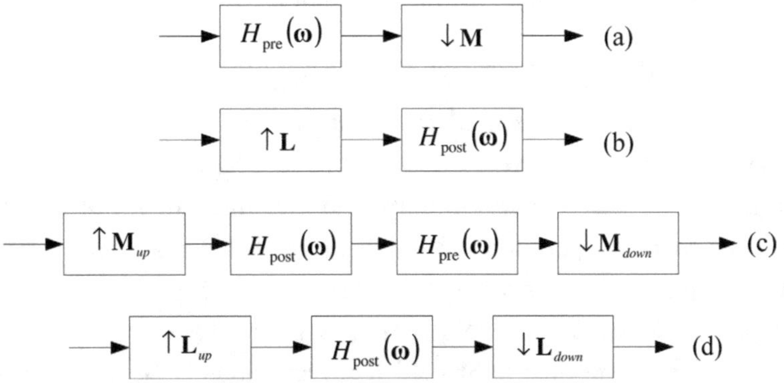

FIGURE 3.2: Discrete domain down-sample and up-sample filters. (a) Down-sampling. (b) Up-sampling. (c) Down-sampling with rational matrix. (d) Up-sampling with rational matrix

where $2\pi\mathbf{k}$ is there to emphasize the periodicity of the spectrum, which can be ignored if we only concern ω in $[-\pi, \pi)^D$.

For a nonsingular integer matrix $\mathbf{L}$, the $\mathbf{L}$-fold up-sampling of $l_d(\mathbf{n})$ is defined as [11]

$$r_d(\mathbf{n}) = \begin{cases} l_d(\mathbf{L}^{-1}\mathbf{n}) & \mathbf{n} \in \text{LAT}(\mathbf{L}) \\ 0 & \text{otherwise,} \end{cases} \qquad (3.9)$$

where $\text{LAT}(\mathbf{L})$ (the lattice generated by $\mathbf{L}$) denotes the set of all vectors of the form $\mathbf{Lm}, \mathbf{m} \in \mathcal{N}$. Clearly, the condition $\mathbf{n} \in \text{LAT}(\mathbf{L})$ above is equivalent to $\mathbf{L}^{-1}\mathbf{n} \in \mathcal{N}$. The corresponding frequency domain relationship of up-sampling is

$$R_d(\omega) = L_d(\mathbf{L}^T\omega). \qquad (3.10)$$

Similarly, up-sampling is normally followed by postfiltering to fill in the numbers that are padded as zeros, as shown in Fig. 3.2(b).

The above down-sampling and up-sampling of discrete signals are rather limited, because both the down-sampling matrix $\mathbf{M}$ and the up-sampling matrix $\mathbf{L}$ can only contain integers. In general, however, we can cascade two down-sampling and up-sampling operations to produce rational sampling matrices. For instance, let $\mathbf{M}$ be a rational matrix for down-sampling. We can always decompose $\mathbf{M}$ as the multiplication of two integer matrices,

$$\mathbf{M} = \mathbf{M}_{\text{up}}^{-1} \cdot \mathbf{M}_{\text{down}}. \qquad (3.11)$$

Hence the down-sampling process can be realized through Figure 3.2(c), where we first up-sample the signal with matrix $\mathbf{M}_{\text{up}}$, then pass it through an up-sampling postfilter, followed by a down-sampling prefilter, and finally down-sample the signal with matrix $\mathbf{M}_{\text{down}}$. The overall effect is that the signal is down-sampled with rational matrix $\mathbf{M}$. Rational matrix up-sampling can be implemented in a similar way, as shown in Fig. 3.2(d). Note that we always perform up-sampling before down-sampling to avoid information loss.

3.2   CONTINUOUS LIGHT FIELD SAMPLING

In the previous chapter, we have derived the spectrum of a light field. If no occlusion and non-Lambertian effects are considered, the spectrum of a light field has a fan shape and lies between two lines determined by the minimum and maximum depth of the scene, as shown in Fig. 3.3(a). Note that the spectrum is in high-dimensional space and there are unlimited ways to sample the signal as long as the replicated spectrums (Fig. 3.1(b)) do not overlap each other. In this book, we focus on two major sampling strategies—rectangular sampling which has been widely used in traditional light field representations [29], as shown in Fig. 3.3(b); and a particular nonrectangular sampling scheme which is theoretically optimal, as shown in Fig. 3.3(c).

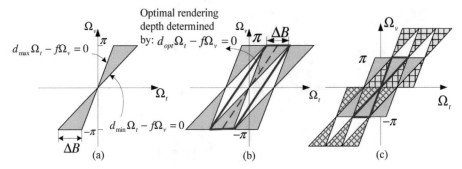

FIGURE 3.3: Rectangular and nonrectangular sampling of light field. (a) Spectrum of a light field assuming no occlusion and Lambertian surface. (b) The optimal rectangular compacting and the corresponding reconstruction filter. (c) The most compact way to pack the light field spectrum with generalized sampling

3.2.1   Rectangular Sampling

Fig. 3.3(b) shows the optimal rectangular sampling scheme one can achieve with rectangular sampling. The minimum distance between the replicas of the spectrums is

$$\Delta B = \pi f\left(\frac{1}{d_{\min}} - \frac{1}{d_{\max}}\right),\tag{3.12}$$

where f is the focal length, and $d_{\min}$ and $d_{\max}$ are the minimum and maximum depths of the scene. The maximum sampling distance along the x_t camera line is thus

$$\Delta x_t = \frac{2\pi}{\Delta B}.\tag{3.13}$$

The reconstruction filter for the above sampling strategy is also given in Fig. 3.3(b). It is a tilted parallelogram, as marked with bold contours. The explicit form of the ideal reconstruction filter in the original light field domain can be written as

$$f_r(\mathbf{x}) = \frac{\sin\left(\frac{\Delta B x_t}{2}\right)}{\frac{\Delta B x_t}{2}} \cdot \frac{\sin \pi \left(x_v + \frac{f x_t}{d_{\text{opt}}}\right)}{\pi \left(x_v + \frac{f x_t}{d_{\text{opt}}}\right)},\tag{3.14}$$

where

$$d_{\text{opt}} = \frac{2}{\frac{1}{d_{\max}} + \frac{1}{d_{\min}}}.\tag{3.15}$$

The ideal reconstruction filter is the product of two sine functions along two nonorthogonal directions, which we call the *principle directions* of the interpolation filter.

In practice, the ideal reconstruction filter has to be truncated for efficient computation, and bilinear interpolation is often the most preferred method for reconstruction due to its

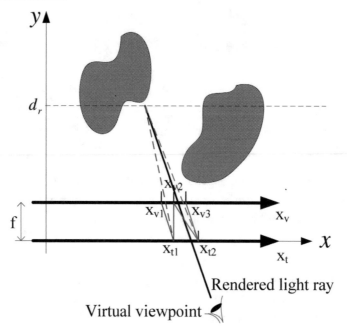

FIGURE 3.4: Depth-corrected bilinear interpolation for rectangular sampling of light field

simplicity and effectiveness. As shown in Fig. 3.4, a to-be-rendered light ray crosses the camera line and the focal line at (t, v). The newest sample points on the camera line are x_{t1} and x_{t2}, and the ones on the focal line are x_{v2} and x_{v3}. Naïvely, light rays (x_{t1}, x_{v2}), (x_{t1}, x_{v3}), (x_{t2}, x_{v2}), and (x_{t2}, x_{v3}) shall be used to preform bilinear interpolation and obtain the rendered light ray. However, if we know that the scene is roughly at depth d_r, it is advantageous to use the light rays (x_{t1}, x_{v1}) and (x_{t1}, x_{v2}) from camera x_{t1} instead, as shown in Fig. 3.4. This depth-compensated light field rendering technique was first discovered in [17], and has proven to be very important in improving the rendering quality of the light filed. From Fig. 3.3(b), it is easy to tell that if a single depth is used for rendering a scene between d_{min} and d_{max}, the virtual rendering depth should be set to d_{opt}.

When the scene contains highly reflective surfaces, as discussed in the previous chapter, the light field spectrum will be expanded. Accordingly, the minimum sampling rate along the camera line shall be reduced, as shown in Fig. 3.5. The minimum distance between spectrum replicas becomes

$$\Delta B = \pi f \left(\frac{1}{d_{min}} - \frac{1}{d_{max}} \right) + 2 B^E, \qquad (3.16)$$

where B^E can be estimated according to Eq. (2.32). The optimal rendering depth stays as d_{opt} in Eq. (3.15).

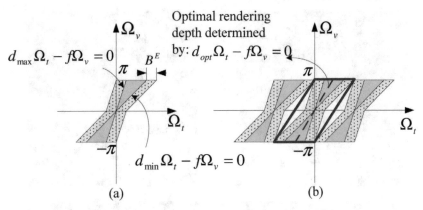

FIGURE 3.5: Rectangular light field sampling for non-Lambertian scenes

In practice, the effect of the non-Lambertian property in expanding the spectrum is often difficult to observe because the light rays from the same surface point change radiance often very slowly. We next show an example where we have to take more images in order to render the scene correctly because of the non-Lambertian property. We render two spheres (Figs. 3.6(a) and 3.6(b)), one is a Lambertian and the other is a purely reflective surface. We found that at a certain sampling rate, we are able to reconstruct the Lambertian scene very well as in Fig. 3.6(c). However, if we use the same sampling rate for the reflective surface, the reconstruction is very bad (Fig. 3.6(d)). This means that we need to sample more images to reconstruct the latter surface. In Figs. 3.6(e) and 3.6(f), we show the EPI of both scenes (center row, at a much higher sampling rate along t for illustration purpose). The red dashed lines are interpolation directions determined by the sphere geometry. Obviously, the rendering or interpolation of the reflective surface is along a wrong direction, or using a wrong geometry, which causes its poor quality.

When occlusions dominate in the scene, the optimal rectangular sampling strategy also needs to be revised. Fig. 3.7 illustrates the idea. In the worst case, the major occluding objects are at d_{min}, which generates additional modulated components along the slope $\frac{d_{min}}{f}$, where f is the focal length. The corresponding spectrum is given in Fig. 3.7(a). If we still apply the rectangular sampling strategy, the spectrum can be compacted as in Fig. 3.7(b). Note that we have to increase the minimum sampling rate by a factor of 2. In addition, from Fig. 3.7(b) we see that the optimal rendering depth becomes d_{min} instead of d_{opt} in Eq. (3.15). In general, when the major occluding objects are at distance d_{occ}, the spectrum is shown in Fig. 3.7(c). After the most compact rectangular sampling, the spectrum is shown in Fig. 3.7(d). The minimum sampling rate and the optimal rendering depth both depend on d_{occ}.

We give an example to verify the above analysis. In Fig. 3.8 we show an OpenGL scene composed of $3 \times 3 \times 3$ cubes. The cubes stay on a 3D regular grid and the front cubes occlude most regions of the back cubes. The renderings are done through depth-corrected bilinear

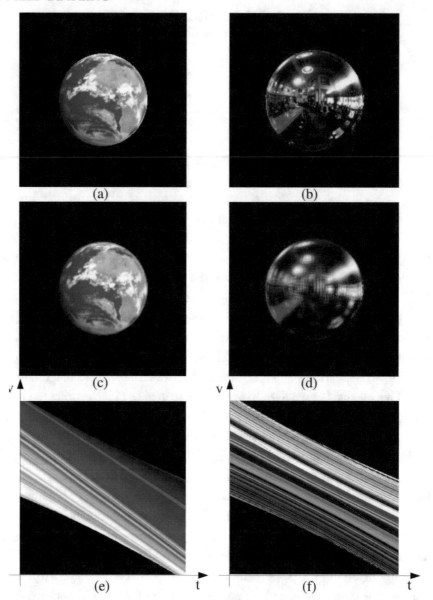

FIGURE 3.6: An example of the minimum sampling rate for non-Lambertian surface. (a) A Lambertian sphere; (b) a purely reflective sphere; (c) at a certain sampling rate, we are able to reconstruct the scene very well for the Lambertian sphere; (d) at the same sampling rate, the reconstruction is very bad for the reflective surface; (e) EPI of the Lambertian sphere scene (center row); (f) EPI of the reflective surface scene (center row)

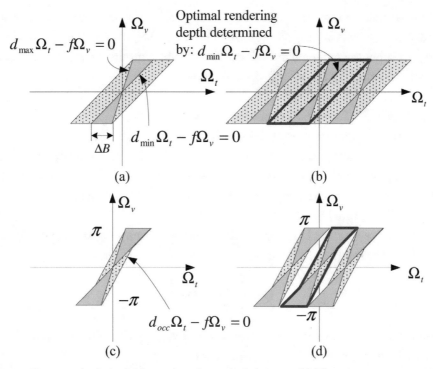

FIGURE 3.7: Rectangular light field sampling for occluded scenes. (a) The most conservative estimation of the spectrum when the object at d_{min} causes the major occlusions; (b) the optimal rectangular sampling strategy for compacting (a); (c) the spectrum given that the object at d_{occ} is the main source of occlusions; (d) the optimal rectangular sampling strategy for compacting (c)

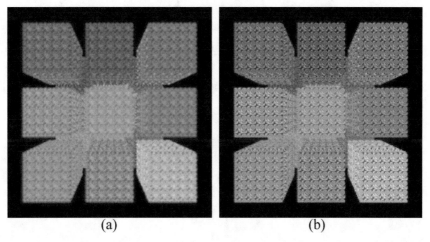

FIGURE 3.8: An example of the minimum sampling rate for occluded surface. (a) At a certain sampling rate, rendered with d_{opt}; (b) same as (a) but rendered with d_{min}

interpolation assuming a constant depth. Figs. 3.8(a) and 3.8(b) show the rendered scene with rendering depth d_{opt} and d_{min}, respectively. Obviously, Fig. 3.8(b) is much more pleasing because the foreground cubes are rendered at high quality. The background cubes are occluded and do not contribute too much to the overall visual quality.

3.2.2   Theoretically Optimal Nonrectangular Sampling

Given the spectrum support of light field, it is possible to compact the replicas better with generalized sampling theory, as shown in Fig. 3.3(c). With this theoretically optimal sampling strategy, the sampling efficiency can be improved by a factor of 2 compared to Fig. 3.3(b), which means we only need 50% of the samples. The reconstruction filter is marked in bold contour in Fig. 3.3(a), which is a tilted fan-like filter.

The optimal replication lattice $\mathbf{U}$ for the generalized sampling strategy in the frequency domain can be easily found as follows:

$$\mathbf{U} = \begin{bmatrix} \frac{\pi f}{d_{min}} & \frac{\pi f}{d_{max}} \\ \pi & \pi \end{bmatrix}. \tag{3.17}$$

In the spatial domain, the above replication in the frequency domain corresponds to a parallelepiped lattice sampling structure with sampling matrix $\mathbf{V}$,

$$\mathbf{V} = 2\pi \mathbf{U}^{-T} = 2d_e \begin{bmatrix} \frac{1}{f} & -\frac{1}{f} \\ -\frac{1}{d_{max}} & \frac{1}{d_{min}} \end{bmatrix}, \tag{3.18}$$

where d_e is defined by

$$\frac{1}{d_e} = \frac{1}{d_{min}} - \frac{1}{d_{max}}. \tag{3.19}$$

To show what happens in the spatial domain when we sample with the sampling matrix given above, we show an example in Fig. 3.9. The circles represent the data samples we keep after the sampling. The lattice is derived by assuming $d_{min} = \frac{4}{3}f$ and $d_{max} = 4f$. Therefore, our sampling matrix is

$$\mathbf{V} = 2\pi \mathbf{U}^{-T} = \begin{bmatrix} 4 & -4 \\ -1 & 3 \end{bmatrix}. \tag{3.20}$$

As a comparison, the rectangular sampling approach has a sampling matrix

$$\mathbf{V} = \begin{bmatrix} 4 & 0 \\ 0 & 1 \end{bmatrix}, \tag{3.21}$$

which is diagonal. Comparing the number of samples per unit area, $\frac{1}{|\mathbf{V}|}$, of the two sampling matrices it is obvious that the new approach saves half of the samples.

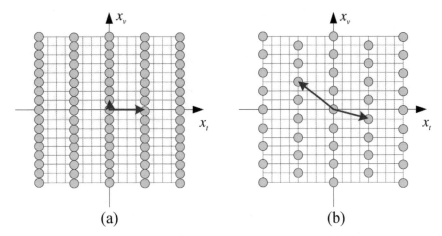

FIGURE 3.9: Example sampling lattices of (a) rectangular sampling and (b) theoretically optimal nonrectangular sampling

The reconstruction filter of the above theoretically optimal nonrectangular sampling is shown as the bold contour in Fig. 3.3(c) in the frequency domain. Although the explicit form of the tilted fan-like filter is available, it will have to be adjusted for each scene as the depth range may vary. A better way to reconstruct the continuous scene is to first take a transform from the $\mathbf{x}$ space to another space $\mathbf{y} = [y_1\ y_2]^T$ through

$$\mathbf{y} = \mathbf{V}^{-1}\mathbf{x}. \tag{3.22}$$

After the transform, sampling in the $\mathbf{x}$ space by sampling matrix $\mathbf{V}$ is equivalent to sampling in the $\mathbf{y}$ space by a 2×2 identity matrix, and the sampling index is given by the integer vector $\mathbf{n}$ in Eq. (3.2). Hence we can obtain the continuous signal over $\mathbf{y}$ first and then transform it back to the $\mathbf{x}$ domain.

The Fourier transform of the sampled signal in the $\mathbf{y}$ space is shown in Fig. 3.10(a). The reconstruction filter is drawn in bold in Fig. 3.10(a) and its spatial domain response is shown in Fig. 3.10(b). The explicit form of the spatial domain reconstruction filter is

$$f_r(\mathbf{y}) = \frac{y_2 \sin^2(\pi y_1) - y_1 \sin^2(\pi y_2)}{\pi^2 y_1 y_2 (y_1 - y_2)}. \tag{3.23}$$

Interestingly, if we look at the filter values along the direction where $y_1 + y_2 = 0$, the filter can be written as:

$$f_r(\mathbf{y})_{y_1 + y_2 = 0} = \frac{\sin^2(\pi y_1)}{\pi^2 y_1^2}, \tag{3.24}$$

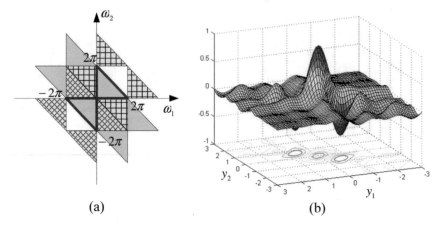

FIGURE 3.10: Reconstruction filter for theoretically optimal sampling. (a) The frequency support and optimal compacting in the transformed frequency space. (b) The standard reconstruction filter

which is a low pass filter. On the other hand, along the direction where $y_1 - y_2 = 0$, this filter can be represented as

$$\lim_{y_1 - y_2 \to 0} f_r(\mathbf{y}) = \frac{\sin \pi y_1 \left(2\pi y_1 \cos(\pi y_1) - \sin(\pi y_1)\right)}{\pi^2 y_1^2}, \qquad (3.25)$$

which is a high pass filter and has a minimum peak of value -0.587 around $y_1 = y_2 = \pm 0.65$. In practice, both the reconstruction filter for the previous approach in [6] and the one for the proposed approach have infinite support. Truncation or other techniques have to be employed to roughly reconstruct the original continuous signal.

Nonrectangular sampling can certainly be applied to scenes with non-Lambertian surfaces and heavy occlusions. Fig. 3.11 shows a number of ways to pack the spectrums in the frequency

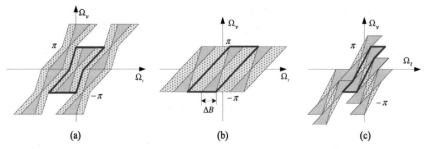

FIGURE 3.11: Theoretically optimal sampling for non-Lambertian and occluded scenes. (a) Non-Lambertian scene. (b) Occluded scene with the object at minimum distance being the dominant factor. (c) Occluded scene with the object at certain distance being the dominant factor

domain for the non-Lambertian and occluded scenes. Note that the packing scheme heavily depends on the scene property.

3.3   DISCRETE LIGHT FIELD SAMPLING

The theoretically optimal nonrectangular light field sampling analysis in the continuous domain is sound in theory, but it is difficult to implement directly in practice. Take the example given in Fig. 3.9(b) as an example. When the capturing camera moves along the camera line x_t, the pixel locations of each camera need to be offset to achieve the optimal sampling. A more feasible approach is to first take as many images for the scene as possible using rectangular sampling and then analyze the captured data and try to remove the redundant pixels that have been taken. This leads to the analysis of light field sampling in the discrete domain.

Assume that we are given a set of images captured using rectangular sampling matrix $\mathbf{V}$, where $\mathbf{V}$ is diagonal. The distance between neighboring cameras is small enough so that there is no aliasing effect between the replicated spectrums. The discrete-time Fourier transform of the discrete signal is shown in Fig. 3.12(a). The frequency response is still bounded by two lines passing through the origin, with slopes $k_{\min}$ and $k_{\max}$, similar to the original continuous spectrum. However, since the distance between neighboring cameras may not be the same as the distance between neighboring pixels in the captured images, $k_{\min} = \frac{d_{\min}}{f}$ and $k_{\max} = \frac{d_{\max}}{f}$ do not necessarily hold. Nevertheless, $k_{\min}$ and $k_{\max}$ can be easily obtained through the Fourier analysis of the original discrete signal (or through the sampling matrix $\mathbf{V}$) and we do not care about what $d_{\min}$, $d_{\max}$, and f really are in the following analysis.

Given the oversampled discrete light field, discrete down-sampling can be applied to reduce the number of samples. Fig. 3.12(b) shows the rectangular down-sampling, whose

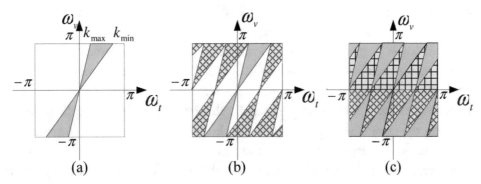

FIGURE 3.12: Discrete light field sampling. (a) Fourier transform of the oversampled discrete signal. (b) The Fourier transform of the signal after rectangular down-sampling. (c) The Fourier transform of the signal after theoretically optimal down-sampling

spectrum duplication matrix is

$$\mathbf{N} = \begin{bmatrix} \frac{\pi}{k_e} & 0 \\ 0 & 2\pi \end{bmatrix}, \tag{3.26}$$

where k_e is defined by

$$\frac{1}{k_e} = \frac{1}{k_{\min}} - \frac{1}{k_{\max}}. \tag{3.27}$$

And the sampling matrix is

$$\mathbf{M} = 2\pi \mathbf{N}^{-T} = \begin{bmatrix} 2k_e & 0 \\ 0 & 1 \end{bmatrix}. \tag{3.28}$$

For comparison, using the optimal nonrectangular sampling scheme, we have

$$\mathbf{N} = \begin{bmatrix} \frac{\pi}{k_{\max}} & \frac{\pi}{k_{\min}} \\ \pi & \pi \end{bmatrix}. \tag{3.29}$$

And the sample matrix is:

$$\mathbf{M} = 2\pi \mathbf{N}^{-T} = 2k_e \begin{bmatrix} 1 & -1 \\ -\frac{1}{k_{\max}} & \frac{1}{k_{\min}} \end{bmatrix}. \tag{3.30}$$

Note the similarity between Eqs. (3.30) and (3.18). The above sampling matrices may not be integer matrices. However, we can always approximate them with a rational matrix and employ the sampling scheme introduced in Section 3.1.2. We refer the reader to [11] for a more detailed explanation on rational matrix down-sampling.

Reconstruction from the discretely down-sampled signal is identical to that from the continuous domain sampled signal discussed in Section 3.2.2. In practice, such a reconstruction filter has to have a limited size of support for a reasonable speed of rendering. In [77], a reconstruction filter with limited support for the theoretically optimal sampling scheme was designed using the eigenfilter approach [68] and showed slightly better performance over the bilinear filter under rectangular sampling and same sampling rate for scenes with a lot of high-frequency components. For a regular light field, which often has weak high-frequency components and strong low-frequency components, the two methods are similar in accuracy but the latter has much lower computational complexity. Rectangular sampling is thus more preferable in practice despite its suboptimal characteristics in theory.

3.4   SAMPLING IN THE JOINT IMAGE AND GEOMETRY SPACE

It is conceptually well known that if the scene geometry is partially known, the number of images needed for perfect light field rendering can be reduced. However, it is not clear to what extent the additional depth information can help, because the joint image and geometry space is usually difficult to imagine and analyze. In this section, we present some important quantitative results for the sampling problem in the joint image and geometry space based on the work in [8].

Let us take a second look at the spectral analysis approach for scenes with unknown geometry discussed in Section 2.3. There we assumed that the scene can be approximated using truncated and piecewise constant depth segments $d_j, 0 \leq j \leq J - 1$. We concluded that the whole scene's spectrum can be approximated as the sum of Fourier transforms for all depth layers; hence, the overall spectrum is bounded by two lines determined by the minimum and maximum depth of the scene, $d_{\min}$ and $d_{\max}$. As shown in Fig. 3.13(a), the different depth layers divides the full spectrum of the scene into J segments, each representing a single layer. Since during the rendering process we may render different scene layers using their corresponding depth values, the sampling rate of such a scene can be considered layer by layer. In the optimal case, the J depth layers divide the bandwidth ΔB (defined in Eq. (3.12)) equally; hence the bandwidth of each layer is

$$\Delta B' = \frac{\Delta B}{J}. \tag{3.31}$$

The new maximum sampling distance along x_t is

$$\Delta x_t' = \frac{2\pi}{\Delta B'} = \frac{2\pi J}{\Delta B}. \tag{3.32}$$

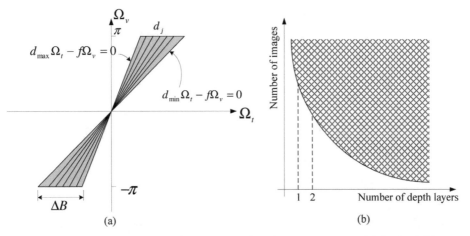

(a)   (b)

FIGURE 3.13: Light field sampling in the joint image and geometry space. (a) Light field spectrum in multiple layers. (b) The minimum sampling curve

Note that the distance is proportional to the number of depth layers. In other words, the number of images needed or the sampling rate is inversely proportional to the number of depth layers. This is shown in Fig. 3.13(b). The curve is named "the minimum sampling curve" in [8]. The shadow region above the curve indicates sampling points that are redundant, while the region below the curve indicates sampling points that are insufficient and can cause aliasing. Note that when ΔB varies, the minimum sampling curve will move up and down according to Eq. (3.32); hence there are a family of such curves corresponding to difference scenes.

In general, when the J depth layers do not divide the bandwidth ΔB equally, or when there is uncertainty about the depth information, we assume that each layer covers a bandwidth of ΔB_i, $i = 0 \leq j \leq J - 1$; the maximum sampling distance along x_t shall be

$$\Delta x_t'' = \frac{2\pi}{\max_j \Delta B_j}. \tag{3.33}$$

The above quantitative analysis of light field sampling in the joint image and depth space has a number of potential applications. For instance, given the number of image samples one can afford, the minimum sampling curve determines how much depth information is needed and vise versa. In image-based rendering it is quite often that the scene geometry is first reconstructed from a set of captured images. The above analysis provides guidance on how much accuracy the depth should be recovered in order to achieve great rendering quality. Rendering-driven depth reconstruction can be very different from the traditional geometry-driven depth reconstruction, as the purpose of the former is to have good rendering quality, while the purpose of the latter is to recover good geometry.

CHAPTER 4

The Freeform Sampling Framework

In the previous chapters, we studied the uniform sampling of the light field data using the traditional Fourier transform based sampling analysis method. Such analysis may provide a general guidance on how a scene should be sampled. For instance, if a scene has non-Lambertian surface or occluded regions, more samples are needed. On the other hand, for many real-world scenes, constraining the samples to be uniformly distributed already limits the best sampling efficiency one can achieve. As a simple example, consider a real-world scene that is composed of both the Lambertian and non-Lambertian surfaces. The Lambertian surface can be sampled at a relatively low sampling rate, while the non-Lambertian surface may need a high sampling rate. Uniformly sampling the scene without being concerned about the regional surface property may easily cause over-sampling of the Lambertian surface or under-sampling of the non-Lambertian surface.

In this chapter, we describe a new sampling framework called *freeform sampling*. In contrast to the traditional uniform sampling approach, where samples have to be taken according to a fixed sampling pattern, freeform sampling allows the sample locations to be irregular or nonuniform. Freedom sampling has the potential to greatly outperform the uniform sampling efficiency, hence allowing a signal to be reconstructed from much fewer samples.

4.1 PROBLEM STATEMENT

Consider a function in the form of $y = f(\mathbf{x})$, $\mathbf{x} \in \mathbf{X}$, where $\mathbf{X}$ is the region of support (ROS). Here $\mathbf{x}$ can be a variable, a vector, a matrix, or anything else that is meaningful. For example, in the scenario of light field sampling, y is the light ray captured at $\mathbf{x}$, where $\mathbf{x}$ is the 4D vector (x_s, x_t, x_u, x_v) that uniquely parameterizes the light ray.

To find out what the unknown function $f(\mathbf{x})$ is, one may take a set of samples in the ROS of $\mathbf{X}$. Let the *sample set* be $\mathbf{X}^s \subset \mathbf{X}$. In freeform sampling, this sample set can be irregularly distributed. The *sample set values*, denoted as $\tilde{f}(\mathbf{X}^s)$, may subject to some noise $n(\mathbf{X}^s)$,

$$\tilde{f}(\mathbf{X}^s) = f(\mathbf{X}^s) + n(\mathbf{X}^s). \qquad (4.1)$$

Given such a sample set and its values, the original signal can be reconstructed. For any given $\mathbf{x} \in \mathbf{X}$, we denote the reconstructed signal value as

$$\hat{f}\left(\mathbf{x}\middle|(\mathbf{X}^s, \tilde{f}(\mathbf{X}^s)), \mathcal{R}\right), \tag{4.2}$$

where $\mathcal{R}$ is the reconstruction method. Note that when we choose different reconstruction methods, the above reconstruction may present different results. In many cases, the reconstructed signal may be different from the original function, thus an error function $e(\mathbf{x})$, $\mathbf{x} \in \mathbf{X}$ can be defined. The concrete form of the error function is user or application dependent. For example, one possible error function is the squared error,

$$e(\mathbf{x}) = \left\| f(\mathbf{x}) - \hat{f}\left(\mathbf{x}\middle|(\mathbf{X}^s, \tilde{f}(\mathbf{X}^s)), \mathcal{R}\right) \right\|^2. \tag{4.3}$$

The general freeform sampling problem can thus be defined as:

Definition 4.1.1. Freeform sampling: *Given a function* $y = f(\mathbf{x})$, $\mathbf{x} \in \mathbf{X}$, *which is either known or unknown, find a set of sample locations* $\mathbf{X}^s \subset \mathbf{X}$, *such that the reconstruction of y from the sample set* $\mathbf{X}^s$ *using method* $\mathcal{R}$ *meets certain error requirement* $\mathcal{Q}(e(\mathbf{x}))$ *on a reconstruction set* $\mathbf{X}^r \subseteq \mathbf{X}$.

$\mathbf{X}^r$ can be the whole ROS $\mathbf{X}$ or a subset of it. This error requirement is again user or application dependent. For instance, we may let $\mathcal{Q}(e(\mathbf{x}))$ be

$$e\left(\mathbf{x}\middle|(\mathbf{X}^s, \tilde{f}(\mathbf{X}^s)), \mathcal{R}\right) < \varepsilon, \qquad \forall \mathbf{x} \in \mathbf{X}^r, \tag{4.4}$$

in which case we know that the reconstruction error in the reconstruction sample set should be bounded by ε.

The freeform sampling defined above is very general. The traditional Fourier transform based uniform sampling analysis can be consider a special case of freeform sampling, where the sample set $\mathbf{X}^s$ is uniformly distributed, the reconstruction set $\mathbf{X}^r$ is the whole set $\mathbf{X}$, the reconstruction method $\mathcal{R}$ is an interpolation filter designed based on the signal's spectrum, and the error requirement is perfect reconstruction.

On the other hand, freeform sampling differs from uniform sampling in a number of ways. First and most importantly, the sample set $\mathbf{X}^s$ is in freeform instead of being uniformly distributed. Second, we consider the reconstruction method as an important factor which will affect the sampling process, because different methods may produce different reconstructions, and in practice there are a lot of constraints on what reconstruction algorithm one can choose. Third, when we take the sample, there is a noise term. In practice, observations always have noises, and these noises cannot be easily eliminated. Lastly, we introduce the reconstruction sample set $\mathbf{X}^r$. If $\mathbf{X}^r \neq \mathbf{X}$, the best sampling strategy can also be set $\mathbf{X}^r$ dependent. This inspires our view-dependent freeform sampling of the light field discussed in the next chapter.

Another important thing to note is that, given the reconstruction method, since infinitely many functions can have the same value on a sample set $\mathbf{X}^s$ and meet the error requirement, the above sampling problem is meaningful only by adding constraints on $f(\mathbf{x})$. In the traditional uniform sampling theory, a common constraint is that $f(\mathbf{x})$ has band-limited Fourier spectrum [67, 2]. That is, the Fourier transform of $f(\mathbf{x})$ is nonzero only on a finite support. As Fourier transform does not maintain any local information in the spacial domain, such a constraint is not suitable for freeform sampling. Instead, we propose to use local constraints to limit the solution space. More specifically, we constrain that

Condition 4.1.2. $\forall \varepsilon, \mathbf{x} \in \mathbf{X}^r, \exists \delta > 0$, *such that as long as* $d\big(\mathbf{x}, \Phi_N(\mathbf{x})\big) < \delta$, $\mathcal{Q}\big(e(\mathbf{x})\big)$ *as defined in Eq. (4.4) is satisfied.*

Here $\Phi_N(\mathbf{x}) \subset \mathbf{X}^s$ is the nearest N neighbors of $\mathbf{x}$ in $\mathbf{X}^s$. $d\big(\mathbf{x}, \Phi_N(\mathbf{x})\big)$ is the maximum distance from $\mathbf{x}$ to these neighbors,

$$d\big(\mathbf{x}, \Phi_N(\mathbf{x})\big) = \max_{\mathbf{x}_i^s \in \Phi_N(\mathbf{x})} \|\mathbf{x} - \mathbf{x}_i^s\|. \tag{4.5}$$

The neighborhood size N is often application dependent. The above condition is illustrated with a 1D example in Fig. 4.1. In this example $N = 3$. Condition 4.1.2 states that as long as the maximum distance from a reconstruction sample x to its three closest neighboring samples is less than a threshold δ, the reconstructed function value is guaranteed to have an error less than ε.

The above condition implies that the local variation of function $f(\mathbf{x})$ is limited. In fact, if there is no sampling noise, ε is arbitrarily small, and the reconstruction filter is through nearest neighbor interpolation,

$$\hat{f}(\mathbf{x}) = \tilde{f}\big(\Phi_1(\mathbf{x})\big), \tag{4.6}$$

the condition is identical to forcing $f(\mathbf{x})$ to be continuous. In practice, however, we know we will never be able to achieve an error requirement with $\varepsilon \to 0$ due to sampling noises and all

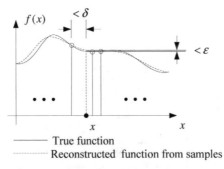

FIGURE 4.1: Illustration of Condition 4.1.2 with a 1D example

sorts of constraints. Hence, we are more interested in cases where ε is acceptably small, and there exists a δ that is reasonably large, which means the error requirement can be met with a finite number of samples.

4.2   PROBLEM AND SOLUTIONS OF FREEFORM SAMPLING

Depending on the application scenario, there are often three types of problems that are under the general freeform sampling framework. In this section, we define these problems and discuss their general solutions by looking at a number of examples.

4.2.1   Sample Reduction

The sample reduction problem is defined as follows:

Definition 4.2.1.   Sample reduction: *Given a set of samples* $\mathbf{X}^s \subset \mathbf{X}$ *and their corresponding function values* $y = f(\mathbf{x}), \mathbf{x} \in \mathbf{X}^s$, *which meets certain error requirement* $\mathcal{Q}(e(\mathbf{x}))$ *with reconstruction method* $\mathcal{R}$ *on a reconstruction set* $\mathbf{X}^r \subseteq \mathbf{X}$, *sample reduction creates a new sample set* $\mathbf{X}^{ss}, |\mathbf{X}^{ss}| < |\mathbf{X}^s|$, *such that the error requirement is still met on* $\mathbf{X}^r$. *Here* $|\mathbf{A}|$ *counts the number of elements in set* $\mathbf{A}$.

Obviously, the uniform discrete down-sampling scheme discussed in Section 3.3 is an example of sample reduction. Note that the smaller sample set $\mathbf{X}^{ss}$ does not have to be a subset of $\mathbf{X}^s$.

When the function $y = f(\mathbf{x})$ is difficult to analyze, or the sample set $\mathbf{X}^s$ is nonuniform, the discrete down-sampling scheme may not be applied any more. Another feasible solution to sample reduction is through progressive decimation, as shown in Fig. 4.2. Starting with the dense set of samples, progressive decimation first identifies which sample can be removed without having problem with the error requirement. If such a sample exists, it decimates this sample, and continues with the next one. Otherwise, it exits. An example of progressive decimation is

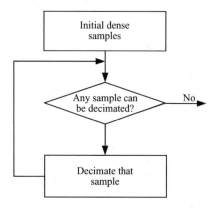

FIGURE 4.2: The flow of progressive decimation for sample reduction

the progressive meshes developed in [19]. An *edge collapse* transformation is used to merge two vertices into one, resulting in a smaller number of vertices in the mesh. The edge collapse transform is progressively applied, such that vertices that can be removed with the least distortions are decimated first. The end result is a representation that can be progressively transmitted, well compressed, and selectively refined. Due to the edge collapse transform, in progressive meshes the smaller sample set $\mathbf{X}^{ss}$ is also not a subset of $\mathbf{X}^s$.

Sample reduction can also be performed assuming $\mathbf{X}^{ss} \subset \mathbf{X}^s$. The best subset $\mathbf{X}^{ss}$ may still require extensive search in general, but there are cases where such search is unnecessary. For instance, in the JPEG image compression standard [47], blocks of images are first transformed by DCT, followed by quantization and entropy coding. Only the DCT coefficients that are nonzero after quantization will be coded into the bitstream and all the other coefficients will be decimated. In this example, the reconstruction set is all the DCT coefficients, and the resampled sample set is the remaining coefficients after quantization (quantization introduces sample noise here). The error requirement is that the difference between the reconstructed DCT coefficients and the original ones is less than the quantization step size. Therefore this is a decremental sampling scheme in the DCT domain. Similar ideas have been widely used in the compression of video [39], light field [81], etc.

Another example of sample reduction is the automatic camera placement algorithm proposed by Fleishman et al. in [13]. Assume that a mesh model of the scene is known; the algorithm strives to find the optimal locations of the cameras such that the captured images can form the best texture map for the mesh model. They found that such a problem can be regarded as a 3D art gallery problem, which is NP-hard [45]. They then proposed an approximation solution for the problem by testing a large set of camera positions and selecting the ones with higher gain rank. The gain was defined based on the portion of the image that can be used for the texture map. The selection process, nevertheless, is a perfect example of sample reduction. Similar approaches include the work by Werner et al. [71], where a set of reference views was selected from a large image pool in order to minimize a target function formulated with the minimum description length principle, and the work by Namboori et al. [42], where an adaptive sample reduction algorithm was developed for a layered depth image [57], which is another representation of image-based rendering.

4.2.2   Minimum Sampling Rate to Meet the Error Requirement

In many applications, we want to know the exact number of samples that can meet the error requirement. The problem is defined as follows:

Definition 4.2.2.   Minimum Sample Rate: *Given a function $y = f(\mathbf{x}), \mathbf{x} \in \mathbf{X}$, find sample set $\mathbf{X}^{ss} \subset \mathbf{X}$, which satisfies $\mathbf{X}^{ss} = \arg\min_{\mathbf{X}^s} |\mathbf{X}^s|$, where $\mathbf{X}^s$ is any sample set that meets error requirement $\mathcal{Q}(e(\mathbf{x}))$ with reconstruction method $\mathcal{R}$ on a reconstruction set $\mathbf{X}^r \subseteq \mathbf{X}$.*

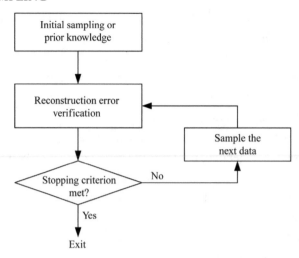

FIGURE 4.3: The flow of incremental sampling

The minimum sampling rate problem can be solved through the sample reduction procedure mentioned above. We can first make a very dense sample of the data and then reduce the sample set to the point that no sample can be further decimated. The drawback is that the initial dense sample set may be expensive or time consuming to acquire.

A popular alternative solution is through incremental sampling, as shown in Fig. 4.3. Incremental sampling starts with an initial sample set, which might be uniform or random following a certain prior probability distribution. This initial sampling process may also be skipped if other prior knowledge is available. Given the set of initial samples or prior knowledge, it then verifies the reconstruction error on the reconstruction sample set. If the error requirement is not met on some of the reconstruction samples, more samples will be taken in the next step to improve the reconstruction of these samples. This process loops until the error requirement is met.

In the recent wavelet-based image compression algorithms such as the SPIHT algorithm [54] and the new JPEG 2000 standard [64], the wavelet coefficients are encoded bitplane by bitplane. Therefore, coefficients with greater magnitudes will be sampled first, and those with smaller magnitudes will be encoded later. This can be considered as an incremental sampling scheme which samples important coefficients first and tries to minimize the energy difference between the original image and the encoded one. A nice property of incremental sampling is that one can stop at any time during the sampling process; thus here the compressed bitstream is embedded or scalable.

The reader may argue that since the wavelet coefficients are known, this example still falls into the category of sample reduction. This is indeed a valid perspective. Incremental sampling becomes more useful when the sampled function $y = f(\mathbf{x})$ is unknown. For instance, Schirmacher et al. [55] proposed an adaptive acquisition scheme for a light field setup. Assuming that the scene geometry is known, they added cameras recursively on the camera plane by predicting the potential improvement in rendering quality when adding a certain view. This *a priori* error estimator accounts for both visibility problems and illumination effects such as specular highlights to some extent. A similar approach was proposed by Zhang and Chen for concentric mosaics setup in [79], where a real system was built to demonstrate the idea. More examples of incremental sampling with unknown function will be presented in Section 4.3.

4.2.3   Minimize Reconstruction Error Given Fixed Number of Samples

In the third category, the total number of samples one can take is assumed to be fixed. The goal is to find the best sample locations such that the reconstruction error is minimized. That is:

Definition 4.2.3.   Minimum Reconstruction Error: *Given a function $y = f(\mathbf{x}), \mathbf{x} \in \mathbf{X}$, find sample set $\mathbf{X}^{ss} \subset \mathbf{X}$, such that $\mathbf{X}^{ss} = \arg\min_{\mathbf{X}^s} \sum_{\mathbf{x} \in \mathbf{X}^r} e(\mathbf{x})$, where $|\mathbf{X}^s| = N$ is any sample set that has fixed size N.*

Note that the error function $e(\mathbf{x})$ is related to the sample set $\mathbf{X}^s$ as Eq. (4.3).

The minimum reconstruction error problem given fixed number of samples is a very complex problem, and a globally optimal solution is often unavailable or NP-hard [15]. Below we describe a rearranged sampling approach that usually achieves a local optimal solution, and we use vector quantization (VQ) as an example of this approach.

The flow of rearranged sampling is shown in Fig. 4.4. Again we start with some initial sampling of the signal. Given the sampled data, the error of the reconstruction sample set is

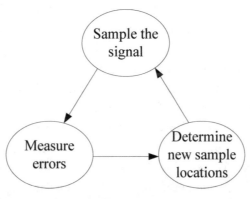

FIGURE 4.4: The flow of rearranged sampling

analyzed. After such analysis, rearranged sampling identifies a new set of sample locations which may possibly reduce the accumulated reconstruction error or cause the error to fit better to the given distribution. It then samples the data at these new locations, which completes a loop. For a static signal, the above loop repeats until the sample locations do not change any more, which results in the most critical set of samples that can be used to represent the whole signal. For a dynamic signal, the above loop repeats until the end of the capturing, which ought to provide better reconstruction quality than a passive sampling approach. Obviously, for dynamic signals, the relocation of the samples should be faster than the signal changes, such that it is reasonable to use the error analysis performed for the last sampling to determine the new sample locations.

Vector quantization (VQ) is a good example of the above rearranged sampling scheme. The goal of VQ is to represent a large set of data points with a small set of codewords. Therefore the codewords form the sample set, and the large data set is the reconstruction set. The reconstruction method is to use the nearest codewords to represent the points in the reconstruction set. The error requirement is that such reconstruction has minimum accumulated error. Furthermore, the popular LBG algorithm [32], which is an iterative algorithm to solve VQ, matches very well with the general flow of the rearranged sampling in Fig. 4.4. One thing to mention is, depending on the initialization, the LBG algorithm may be trapped in local minimum during the iteration, so do all the other practical VQ algorithms. Therefore, rearranged sampling in general may not always achieve the optimal result due to such inherent problems, but it still has a good chance to beat a fixed sampling pattern strategy such as uniform sampling, given the initialization is reasonable. In addition, there are a number of techniques that can partially reduce the chance of a local optimal solution, e.g., simulated annealing [69], genetic algorithms, tabu search [14], etc.

4.3   ACTIVE SAMPLING

In most of the examples we presented in the last section, we know the function values on the reconstruction set. For example, in image compression, the original image (or its transformed version) is the reconstruction set and we know it before sampling. This allows us to easily calculate the reconstruction error given any sample set, and verify if the reconstruction error fulfills the error requirement. Unfortunately, in some other applications, the function values on the reconstruction set is unknown. Therefore, to perform freeform sampling, one must estimate what is the reconstruction error and guess if the error requirement has been fulfilled. Such prediction can only be performed based on samples that have been taken and certain prior knowledge one might have. We term this kind of freeform sampling strategy as *active sampling*. The general solutions of incremental sampling and rearranged sampling is still valid in active sampling. Sample reduction, however, is generally not applicable because we do not have the starting dense sample set.

Despite the difficulty in estimating the reconstruction errors, active sampling has been used in several applications in the literature. There are some shared characteristics of these applications. First, the to-be-sampled function is often unpredictable, thus no fixed sampling pattern can be used to guarantee that the error requirement will be fulfilled, unless a very dense sampling is used. Second, taking a sample in these applications is often very expensive or time consuming. To save the cost or to speed up the sampling process while still fulfilling the error requirement, active sampling becomes the best choice. Below we list some of these application that employs active sampling.

Next best view.   In automated surface acquisition, using range sensors to obtain the object surface is a labor intensive and time-consuming task. Moreover, no fixed sampling pattern can be used because the captured objects often have irregular shape. The main purpose of the next best view (NBV) algorithm is to ensure that all scannable surfaces of an object will be scanned, and determine when to stop scanning [49, 53]. Incremental sampling was used in NBV, where one often determines the next sampling position based on the 3D geometry reconstructed and merged from the previous samples. The error function predicted is usually the holes exhibited in the current geometric model.

Active learning.   For many types of machine learning algorithms, one can find the statistically "optimal" way to select the training data. The pursuing of the "optimal" way by the machine itself was referred to as active learning [9, 26, 18]. For example, a classification machine may determine the next training data as the one that is most difficult for it to classify. Note that in this application, the sample set is the training data, the reconstruction set is the objects in the whole database (or maybe some cross-validation set), and the value of each sample is the label of classification. The error function is predicted as the certainty of the classification. Recently, active learning has been applied to the annotation of database for retrieval [78], which can save a lot of human laboring or annotation.

Motion estimation.   Motion estimation is an important component in modern video compression algorithms [39]. A full search of the motion vector can provide the best motion vectors, but it may be too slow. Various other search algorithms have been proposed, among them the most famous ones are such as the three-step search (TSS) [25] and the four-step search (FSS) [27] algorithms. These searching algorithms first measure the matching error for a subset of possible motion vectors. Only the ones that are promising to be the true motion vector will be refined (more motion vectors will be tested around them and the best one among the tested will be chosen as the final result). This is a good example of sampling–analyzing–sampling loop in incremental sampling.

4.4   ALGORITHMS FOR ACTIVE SAMPLING

When the values of the reconstruction set are known, one can easily measure the reconstruction error, thus the performance of freeform sampling can be largely guaranteed. However, in active sampling such reconstruction error must be estimated. The key to the success of active sampling thus lies in how well we can perform such error prediction.

Since the form of reconstruction error varies from application to application, it is generally not possible to present an algorithm that is suitable to all applications. In this section, we make certain assumptions about the applications and present some algorithms that are widely applicable, particularly for the case of image-based rendering.

4.4.1   Active Incremental Sampling

We first add some structure to the incremental sampling process. Assume that the neighboring samples in the sample set can be organized into cliques, as shown in Fig. 4.5. Here we give three representative cliques: rectangle, triangle, and sample pair. As samples in the same clique are

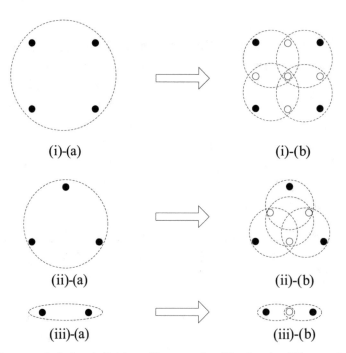

(i)-(a)   (i)-(b)

(ii)-(a)   (ii)-(b)

(iii)-(a)   (iii)-(b)

FIGURE 4.5: Cliques and their subdivision: (i) rectangle; (ii) triangle; (iii) sample pair; (a) original clique; (b) subdivided clique

close to each other, we assume that their sample values should share certain local property. If, for some reason, we believe the local property is not fulfilled, we shall increase the sampling rate by subdividing the clique to smaller ones, as shown in Fig. 4.5(b). This incremental sampling structure naturally leads to nonuniform sampling of the sampled signal. Note that in the above incremental sampling procedure, the local property of the cliques is the key in determining the final sample set. As in active sampling, the final goal is to fulfill the error requirement on the reconstruction set, we shall associate the local property of cliques with the reconstruction method and the reconstruction set.

In many practical applications, it is preferable to have a reconstruction method that is simple, so that the reconstruction can be performed in real time. Here simplicity is two-fold. First, if the function value at a certain location is reconstructed, only a few samples nearby will be involved. Second, the algorithm used for reconstruction is often as simple as some weighted interpolation of the nearby samples. The assumption behind such a reconstruction method is that, albeit a signal can vary violently in large scale, within a small neighborhood it should change slowly.

Under the same assumption above, we hereby claim the following *local consistency principle* for active sampling:

Principle 4.4.1. *Given a to-be-reconstructed sample location, the consistency of the sample values of the nearby samples may serve as a good indicator of its reconstruction error.*

In particular, the consistency of the sample values in a clique may be used to determine whether it should be subdivided or not. Several aspects should be noted in the above principle. First, the concrete form of local consistency between samples may still vary from application to application. Second, the above principle applies only when some to-be-reconstructed sample location is given. If a clique of samples is never used in reconstructing any samples in the reconstruction set, it shall never be subdivided no matter how inconsistent they are. Third, although the local consistency is a good indicator, it cannot be 100% accurate due to the spacing between samples and the sample noise introduced in Eq. (4.1). Therefore, the sample noise and the size of the clique should both be considered while measuring the local consistency. For instance, the accuracy often improves when the samples in the clique get closer.

We therefore detail the flowchart of incremental sampling in Fig. 4.3 with the clique sampling structure and the local consistency principle above in Fig. 4.6. After the initial sampling, we calculate for each clique its local consistency score. Afterwards, we test if any stopping criterion has been met, e.g., if we have reached the specified maximum number of samples, or if all the cliques have scores higher than a certain threshold. If such stopping criterion is not met, we subdivide the clique with the lowest score. This at least helps in reducing the worst-case error if the local consistency reflects the error well enough.

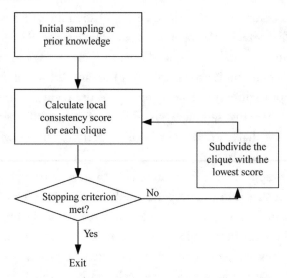

FIGURE 4.6: The flow of active incremental sampling

4.4.2   Active Rearranged Sampling

In rearranged sampling, the number of samples in the sample set is fixed. One must move these samples so that certain error requirement is fulfilled on the reconstruction set. For instance, one may require that the accumulated reconstruction error be minimized, or that the worst reconstruction error be minimized. In both cases, rearranged sampling can be solved using a technique similar to vector quantization.

Assume that at a certain instance, we have captured a set of samples in the sample set. The reconstruction errors of the samples in the reconstruction set can thus be estimated by the local consistency principle. Samples with lower consistency scores may have larger reconstruction error, thus some sample set samples should move toward them to form a denser sampling. In contrast, if some reconstruction set samples have high consistency scores, the sample set samples could be relocated away from them to form a sparser sampling.

The flow of the active rearranged sampling can still be represented by Fig. 4.4, except that the error measurement stage should be replaced by an error estimation stage. Similarly, active rearranged sampling will often use vector quantization like solutions, and we leave the details to the next chapter using the light field as a concrete example.

CHAPTER 5

Light Field Active Sampling

In this chapter, we will apply the active sampling framework in Chapter 4 for the light field. We first discuss the local consistency score of light field applications in Section 5.1. Active incremental sampling and active rearranged sampling are discussed in Sections 5.2 and 5.3, respectively. We show that active sampling outperforms uniform sampling in all the scenes we tested.

5.1   THE LOCAL CONSISTENCY SCORE

The key to active sampling is to define the local consistency score for estimating the reconstruction errors. We mentioned that such consistency score should be related to the reconstruction method, the reconstruction set, the sample noise, and the confidence of applying the consistency measure as the reconstruction error, etc.

The most widely used reconstruction method in light field/image-based rendering (IBR) is through weighted interpolation of nearby captured light rays [29, 58, 5]. In [5], eight goals were proposed for IBR, which led to a weighted interpolation algorithm that considers angular difference, resolution sensitivity, and field of view (FOV). As shown in Fig. 5.1, let OP be a light ray to be rendered. Cameras $C_1, C_2, \ldots, C_K$ are the K nearest cameras to that light ray in terms of their angular difference to OP, which are denoted as $\theta_1, \theta_2, \ldots, \theta_K$. Typically $K = 4$ is good enough. Light rays $C_k P, k = 1, 2, \ldots, K$, will then be used to interpolate OP. We define the weight for angular difference as

$$w_k^{\text{ang}} = \frac{1}{\epsilon + \theta_k}, \qquad k = 1, 2, \ldots, K \qquad (5.1)$$

where ϵ is a small number (e.g., 10^{-10}) to avoid division by zero. If the three-dimensional (3D) point P projects to the image captured by C_k as (x_k, y_k), the weight for FOV $w_k^{\text{fov}}, k = 1, 2, \ldots, K$, is defined as 1 if (x_k, y_k) is inside the FOV, and 0 if it is outside. A narrow region is defined along the FOV boundary to create a smooth transition from 1 to 0. In our discussion, since the capturing cameras lie on a shared plane and we assume that they have roughly the same specifications, the weight for resolution can be assumed as $w_k^{\text{res}} \equiv 1, k = 1, 2, \ldots, K$. The

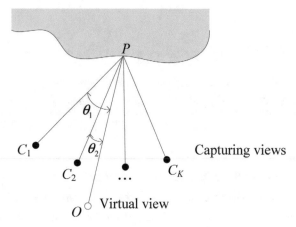

FIGURE 5.1: Interpolation weight calculation using angular difference

overall weight for the light ray $C_k P$ is the multiplication of the three factors:

$$w_k = w_k^{\text{ang}} \times w_k^{\text{fov}} \times w_k^{\text{res}}, \qquad k = 1, 2, \ldots, K. \qquad (5.2)$$

These weights are then normalized to ensure that they sum up to 1 for a given light ray OP. Although our weight definition is different from that in [5], which used a penalty-based weighting scheme, our experiments show that the rendering quality is not sensitive to how the weights are specified, as long as they satisfy the basic requirements such as continuity and epipolar consistency [5].

The local consistency score of the light rays $C_k P, k = 1, 2, \ldots, K$ can be defined in various ways [56, 4]. Let their color intensity be $l_k, k = 1, 2, \ldots, K$. The simplest yet widely used technique is to make use of their intensity variance σ:

$$\mathcal{C} = \frac{1}{\sigma}, \qquad \text{where} \quad \sigma = \sqrt{\frac{1}{K} \sum_{k=1}^{K} (l_k - \bar{l})^2}. \qquad (5.3)$$

Here $\bar{l} = \frac{1}{K} \sum_{k=1}^{K} l_k$ is the average intensity of l_k. The above measure essentially assumes that the surface is Lambertian and not occluded, and reflects the likelihood of the light ray intensities being independent and identical Gaussian distribution as a stochastic variable. On the other hand, under several situations one will obtain a low consistency score: non-Lambertian surface, occluded objects, or inaccurate geometry, as shown in Fig. 5.2. According to the uniform sampling theory in Chapter 2, all these cases in fact require a sampling rate that is higher than normal. In active sampling, regions with a low consistency score imply higher reconstruction error and will be sampled denser; thus it is consistent with the previous conclusions.

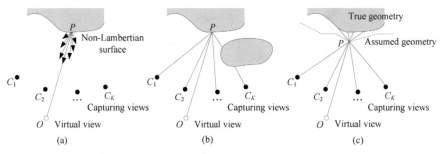

FIGURE 5.2: Cases that will cause a low local consistency score. (a) Non-Lambertian surface; (b) occlusions; (c) inaccurate geometry

If one particular light ray's reconstruction error is considered, such as the light ray OP in Fig. 5.1, we may apply a slightly modified variation assuming the sensor noise is negligible:

$$C = \frac{1}{\sigma'}, \qquad \text{where} \quad \sigma' = \sqrt{\sum_{k=1}^{K} w_k (l_k - \bar{l})^2}. \qquad (5.4)$$

Here $\bar{l} = \sum_{k=1}^{K} w_k l_k$ is the weighted average intensity. Note that w_k should have been normalized to sum to 1. Compared with the standard variance definition in Eq. (5.3), the above weighted variation takes the reconstruction method into consideration, and may better reflect the actual interpolation quality of the given light ray for non-Lambertian or occluded objects.

The confidence of applying the consistency measure as the reconstruction error is hard to integrate into active sampling. It requires certain prior knowledge about the sampled signal. Our solution is to adjust the consistency score obtained in Eqs. (5.3) and (5.4) with a penalty multiplication factor $f(\alpha)$, where α can be the average angular difference $\alpha = \frac{1}{K} \sum_{k=1}^{K} \alpha_k$. The larger the α, the smaller the multiplication factor $f(\alpha)$, the lower the resultant consistency score. $f(\alpha)$ can be considered as a tuning factor which prevents the active sampling from being too aggressive at some local region, thus increasing the robustness of active sampling. It can also be thought of as determining the tradeoff between uniform sampling and active sampling, because if $f(\alpha)$ gives too much penalty for large α, the final sampling result will tend to have the same α everywhere, which becomes a uniform sampling. In the following discussions, we will not worry about the above problem and show only the raw performance of active sampling.

5.2   LIGHT FIELD ACTIVE INCREMENTAL SAMPLING

In the original light field setup [29], the cameras are positioned on a regular grid of the camera plane. For active sampling, we still assume that the cameras are on the camera plane, but they can be arranged nonuniformly. As shown in Fig. 5.3, assume that the capturing cameras

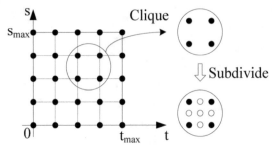

o: Newly captured images during the subdivision

FIGURE 5.3: Active incremental sampling for the light field setup

are within a rectangular range determined by $(0, 0)$ and (s_{max}, t_{max}). We initialize the active capturing process by a reasonably dense uniform sampling. We use the rectangular clique in Fig. 4.5(i) for the subdivision process. That is, every time we capture some new images, we subdivide one of the cliques into four. In the example shown in Fig. 5.3, five new images are taken during the subdivision. It is always that the clique with the lowest accumulated consistency score of all the corresponding light rays (Eq. (5.3)) is subdivided; thus in this example we improve the worst-case rendering quality through active sampling. The sampling process recursively performs the above subdivision until a certain limit on the number of images is reached or the local consistency scores of all the cliques have been smaller than a given threshold.

The above active incremental sampling strategy is tested on a synthetic scene *Earth*, as shown in Fig. 5.4(a). *Earth* is a near-Lambertian scene, whose geometry is known and represented as a $96 \times 96 \times 64$ volumetric model. We initialize the active incremental sampling algorithm by 7×7 uniform sampling, and the overall number of images is limited to be less than or equal to 169. The result is compared to a 13×13 uniform sampling approach. Fig. 5.4(b) shows the final camera arrangements on the camera plane using active incremental sampling; for comparison, (c) shows that of uniform sampling. Each dot represents a camera being there and taking one image. It can be observed that active incremental sampling puts more cameras at the top-right portion of the camera plane. Fig. 5.4(d) is an example view captured in active incremental sampling (red circles in (b)). Uniform sampling did not sample that view. Fig. 5.4(e) is what can be rendered from the sampled images in uniform sampling. As a comparison, Fig. 5.4(f) is a view captured in uniform sampling (red circles in (c)). It is not captured in active incremental sampling but can be rendered as (g). Obviously, we would prefer to sample (d) instead of (f) because the quality degradation from (d) to (e) is more obvious (many high-frequency details are lost in (e)) than that from (f) to (g). Therefore active incremental sampling has taken the right step.

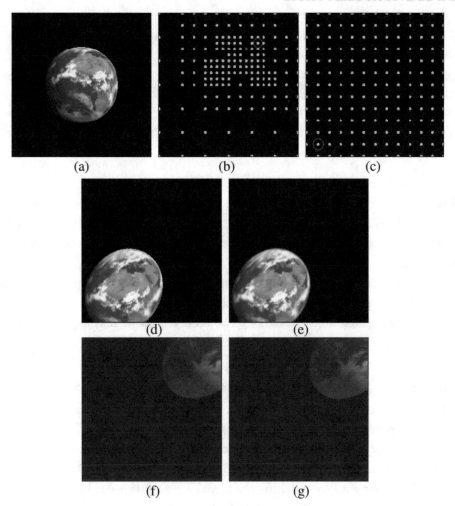

FIGURE 5.4: Active incremental sampling of a light field *Earth*. (a) A snapshot of the *Earth* scene; (b) camera map using active incremental sampling; (c) camera map using uniform sampling; (d) an image captured by active incremental sampling, but not by uniform sampling; (e) rendered image of the same image in (d) from uniform sampling; (f) an image captured by uniform sampling but not by active incremental sampling; (g) rendered image of the same image in (f) from active incremental sampling

To measure the improvement of active incremental sampling over the uniformly sampled light field, we employ two objective measures. The first is the worst-case quality. From the cliques formed by both approaches, we render the virtual views at their centers. As the center views are the farthest from the sampled images, most likely they will have the worst quality. Our first measure is the average peak signal-to-noise ratio (PSNR) of the worst 30 center views. Note that we are able to measure the PSNRs because we are using synthetic scenes and we

TABLE 5.1: Performance Comparison Between Uniform Sampling (US) and Active Incremental Sampling (AIS) on Light Field Scene *Earth*

	US	AIS
Average PSNR of 30 Worst Center Views	32.8 dB	33.2 dB
PSNR Variance of 1000 Rendered Images	3.14 dB	2.61 dB

have the real rendered images as our ground truth. The second measure is the PSNR variance of rendered images. We randomly render 1000 images on the camera plane and measure the variance of the PSNRs. The results are shown in Table 5.1. It can be observed that active IBR has a better worst-case quality and a smaller variance, which is what we expected by performing active sampling.

We may also perform active incremental sampling with triangle cliques, as was shown in Fig. 4.5(ii). The independent work by Schirmacher et al. [55] was such an example, despite some difference in detailed implementation. The same technique may also be applied for the spherical light field [20] if the cliques are defined on the spherical surface.

5.3 LIGHT FIELD ACTIVE REARRANGED SAMPLING

Active rearranged sampling is applicable when the overall number of images one can keep is limited. Here we make a more formal problem statement as follows.

Assume that we have N cameras to capture a static or slowly moving scene. The cameras can move freely on the camera plane, and point to the same direction. During the capturing, we also have P viewers who are watching the scene. These P views are rendered through the method mentioned in Section 5.1 from the N captured images. The goal is to arrange these N cameras such that the P views can be rendered at their best quality. Here both N and P are finite.

Formally, as shown in Fig. 5.5, let the cameras' positions on the camera plane be $\mathbf{c}_j$, $j = 1, 2, \ldots, N$. Since during the rendering, each rendered view will be split into a set of light rays, we combine the P virtual views into L light rays in total. Denote them as l_i, $i = 1, 2, \ldots, L$. For a number of reasons such as insufficient sampling and inaccurate geometric modeling, the rendered light rays are not perfect. Denote the rendering errors of the light rays as e_i, $i = 1, 2, \ldots, L$, or, in vector form, $\mathbf{e}$. Obviously, $\mathbf{e}$ depends on the camera locations $\mathbf{c}_j$. The *camera rearrangement problem* is stated as

Definition 5.3.1. *Given light rays l_i, $i = 1, 2, \ldots, L$, to be rendered, find the optimal camera locations $\hat{\mathbf{c}}_j$, $j = 1, 2, \ldots, N$, such that a function $\Psi(\mathbf{e})$ over the rendering errors is minimized.*

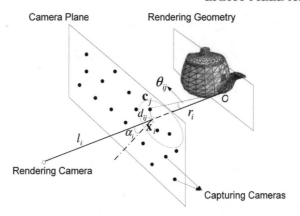

FIGURE 5.5: The formulation of the problem

That is

$$\hat{\mathbf{c}}_j = \arg\min_{\mathbf{c}_j} \Psi(\mathbf{e}).\qquad(5.5)$$

The definition of function $\Psi(\mathbf{e})$ depends on the particular application. In this chapter, we focus on the widely adopted squared error criterion, namely,

$$\Psi(\mathbf{e}) = \sum_{i=1}^{L} e_i^2.\qquad(5.6)$$

We will show, in the next subsections, that such an error function leads to a weighted vector quantization solution for active rearranged capturing, which can be solved efficiently using a modified LBG-VQ algorithm [32].

Note that the above camera rearrangement problem is trivial if all the P virtual views are on the camera plane, and $P \leq N$, because one may simply move the cameras to the virtual view positions and capture the scene at those places directly, leading to zero error for the rendered light rays. However, the problem is nontrivial as long as one of the virtual viewpoints is out of the camera plane (even if $P = 1$), or $P > N$, because any out-of-plane view will have to be synthesized from multiple images, which can be potentially improved by rearranging the capturing cameras.

5.3.1   Formulation Based on the Angular Difference

In Fig. 5.5, denote the intersection of the light rays and the camera plane as $\mathbf{x}_i$, $i = 1, 2, \ldots, L$. Consider a certain light ray l_i, which crosses the scene geometry at O, and one of its neighboring cameras $\mathbf{c}_j$. Denote the distance between $\mathbf{c}_j$ and $\mathbf{x}_i$ as $d_{ij} = \|\mathbf{x}_i - \mathbf{c}_j\|$ and the angular difference as θ_{ij}. Let the distance between O and $\mathbf{x}_i$ be r_i, which is known during the rendering. From the figure, we know that when the scene depth $r_i \gg d_{ij}$ (which is almost always true), we have

$$\theta_{ij} \approx \frac{d_{ij} \cos \alpha_i}{r_i} = w_i \|\mathbf{x}_i - \mathbf{c}_j\| \tag{5.7}$$

where α_i is the angle between the light ray l_i and the normal of the camera plane, and $w_i = \frac{\cos \alpha_i}{r_i}$. Let

$$\widetilde{\theta}_i = \min_{j=1,\dots,N} \theta_{ij} \tag{5.8}$$

be the minimum angular difference between light ray l_i and all the capturing cameras. Intuitively, if the minimum angle $\widetilde{\theta}_i$ is very small, the rendered light ray will be almost aligned with a captured light ray; hence the rendering quality should be high, even if the scene geometry is inaccurate or the scene surface is non-Lambertian. This is also seen in Eq. (5.1). If one of the nearest light rays has a very small angle, its weight will be large and dominate the interpolation process; hence there is little aliasing. The relationship between the rendering error e_i and $\widetilde{\theta}_i$, however, is very complex for practical scenes due to various factors such as geometry accuracy and scene surface property. As a very coarse approximation, we assume

$$e_i = \varepsilon_i(\widetilde{\theta}_i) \approx k_i \widetilde{\theta}_i, \tag{5.9}$$

where k_i is a scaling factor. The right-hand side of Eq. (5.9) is indeed a linear approximation of $\varepsilon_i(\widetilde{\theta}_i)$, which is valid when $\widetilde{\theta}_i$ is very small. The scaling factor k_i, however, differs from light ray to light ray, and is generally unknown.

Active rearranged capturing then minimizes the summation of squared errors as

$$\hat{\mathbf{c}}_j = \arg\min_{\mathbf{c}_j} \Psi(\mathbf{e}) = \arg\min_{\mathbf{c}_j} \sum_{i=1}^{L} e_i^2$$

$$\approx \arg\min_{\mathbf{c}_j} \sum_{i=1}^{L} (k_i \widetilde{\theta}_i)^2$$

$$= \arg\min_{\mathbf{c}_j} \sum_{i=1}^{L} \gamma_i \min_{j=1,\dots,N} \|\mathbf{x}_i - \mathbf{c}_j\|^2, \tag{5.10}$$

where $\gamma_i = w_i^2 k_i^2$. Note that the last equality is due to Eqs. (5.7) and (5.8). The above formulation is a standard weighted vector quantization problem, and can be easily solved if weights γ_i are known. Unfortunately, as mentioned earlier, k_i depends on the geometry accuracy and scene surface property, which is generally unknown.

Although Eq. (5.10) cannot be solved directly, it has some nice properties. For instance, if a certain light ray has a large scaling factor k_i, which means it tends to have a large rendering error, weight γ_i becomes large. The vector quantization process will then reduce more on $\min_{j=1,\dots,N} \|\mathbf{x}_i - \mathbf{c}_j\|^2$, effectively moving the capturing cameras closer to that light ray. Therefore, if weights γ_i can be adjusted according to the rendering quality, the same weighted VQ algorithm can still be used to achieve the best rendering quality. These observations make it clear that we need an iterative solution for active rearranged capturing, which adjusts weights γ_i according to some estimation of the final rendering quality.

5.3.2   A Recursive Algorithm for Active Rearranged Capturing

Fig. 5.6 shows the flowchart of the proposed recursive active rearranged capturing algorithm applicable for static or slowly moving scenes. Given a set of newly captured images, we first estimate the rendering errors of all the light rays. If the viewers have moved (which can cause significant changes to the set of to-be-rendered light rays), weights γ_i in Eq. (5.10) will be reinitialized. Afterward, weights γ_i are updated based on the estimated rendering quality. Weighted VQ is performed as Eq. (5.10), and the cameras are moved to capture the next set of images.

In the above flowchart, the rendering quality, given a set of captured images, can be estimated using the local color consistency, defined in Section 5.1. The remaining problem is how to update weights γ_i. In the following, we present two weight updating mechanisms we experimented on.

The first algorithm is based on the observation that if the weight associated with a certain light ray is increased, the weighted VQ algorithm that follows will tend to move the capturing cameras closer to the light ray. To improve the low-quality light rays, we first define

$$s_i = \log \sigma_i \qquad\qquad (5.11)$$

as the score for each light ray. Let $s_{\min}$ and $s_{\max}$ be the minimum and maximum value of $s_i, i = 1, \ldots, L$. Let $\bar{s}$ be the average value of s_i. Weight γ_i^{t+1} at time instance $t + 1$ is updated

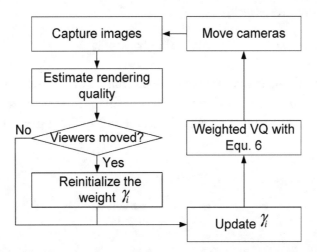

FIGURE 5.6: The flowchart of our proposed active rearranged capturing algorithm for static or slowly moving scenes

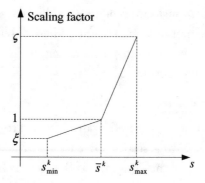

FIGURE 5.7: Scaling factor for updating the auxiliary weights

from those γ_i^t at time instance t as

$$
\gamma_i^{t+1} = \begin{cases} \gamma_i^t * (1 + (\xi - 1)\frac{\bar{s}-s_i}{\bar{s}-s_{\min}}), & s_i \leq \bar{s}, \\ \gamma_i^t * (1 + (\zeta - 1)\frac{s_i-\bar{s}}{s_{\max}-\bar{s}}), & s_i > \bar{s}, \end{cases} \tag{5.12}
$$

where ξ and ζ are the minimum and the maximum weight scaling factor. They are set as 0.5 and 4, respectively, in the current implementation. As illustrated in Fig. 5.7, Eq. 5.12 states that if the variance of the projected pixels for a light ray is greater than the average (thus the local color consistency is bad), its weight will be increased. During the weighted VQ, the cameras will then move closer to that light ray. Otherwise, the cameras will move away. Note that after the weight update with Eq. (5.12), one should normalize the new weights such that $\sum_{i=1}^{L} \gamma_i^{t+1} = 1$.

Eq. (5.12) requires initial values of γ_i^0 to start the iteration. We find in practice that the following initialization works well,

$$
\gamma_i^0 \propto \min_{j=1,\dots,N} \|\mathbf{x}_i - \mathbf{c}_j\|, \tag{5.13}
$$

which gives higher weights to light rays that are far from any capturing cameras.

The second weight updating algorithm is based on the linear approximation between the rendering error and the minimum angular difference in Eq. (5.9); hence it is straightforward that

$$
\begin{aligned}
\gamma_i &= w_i^2 k_i^2 \\
&\approx \frac{e_i^2}{\min\limits_{j=1,\dots,N} \|\mathbf{x}_i - \mathbf{c}_j\|^2} \\
&\approx \frac{\sigma_i^2}{\min\limits_{j=1,\dots,N} \|\mathbf{x}_i - \mathbf{c}_j\|^2}.
\end{aligned} \tag{5.14}
$$

Since the rendering errors e_i are estimated from the local color consistency, the second weight updating algorithm does not require initial values of γ_i. That is, even if the viewers keep moving around, the weight reinitialization step in Fig. 5.6 can be skipped, making it more adaptable for viewer changes and dynamic scenes.

The recursive active rearranged capturing algorithm is thus summarized in Fig. 5.8.

Given a set of images newly captured, perform:

1. **Error estimation**, estimate the rendering error using local color consistency for any light ray l_i:

   Rendered color: $m_i = \sum_{k=1}^{K} \mu_k I_{ik}$
   Variance: $\sigma_i^2 = \sum_{k=1}^{K} \mu_k (I_{ik} - m_i)^2$

   where the summations are for the K nearest neighbors. μ_k is the weight for each captured light ray determined by the rendering algorithm, I_{ik} is the captured light ray intensity.

2. **Weight update**, update the weights γ_i as Eq. (5.12) or (5.14).

3. **Weighted VQ**, perform weighted vector quantization with a modified LBG-VQ algorithm [32]:

   a. **Nearest neighbor condition:**

   $\mathbf{x}_i \in \mathcal{R}_j$, if $\|\mathbf{x}_i - \mathbf{c}_j\| \le \|\mathbf{x}_i - \mathbf{c}_{j'}\|, \forall j' = 1, \cdots, N$

   where $\mathcal{R}_j$ is the neighborhood region of centroid $\mathbf{c}_j$.

   b. **Centroid condition:**

   $\mathbf{c}_j = \dfrac{\sum_{\mathbf{x}_i \in \mathcal{R}_j} \gamma_i \mathbf{x}_i}{\sum_{\mathbf{x}_i \in \mathcal{R}_j} \gamma_i}, j = 1, \cdots, N$

4. **Capture images**, move the cameras according to the VQ result and capture images. (Go back to 1).

FIGURE 5.8: The recursive active rearranged capturing algorithm

5.3.3   Experimental Results

We verify the effectiveness of the proposed view-dependent active rearranged capturing algorithm with three synthetic scenes, namely, *teapot*, *vase*, and *wineglass* (Fig. 5.10), all rendered from POV-Ray [50], which creates 3D photorealistic images using ray tracing. The scenes contain complex textures, occlusions, semireflection, transparency, etc. In all cases, we use 64 cameras on the camera plane to capture the scene, as shown in Fig. 5.9. The initial camera locations are uniformly distributed on the camera plane. The virtual viewpoints can be anywhere for light field rendering, but for experimental purposes we place them on a plane in front of the camera plane, at a distance z_0, where z_0 roughly equals to the distance between neighboring initial camera locations d_0. The virtual viewpoints are assumed to be on a rectangular grid, indexed from 1 to 9, as shown in Fig. 5.9. The distance between neighboring virtual viewpoints is d.

We first consider the case where a single virtual view is rendered from the viewpoint indexed as 1 in Fig. 5.9. We use a constant depth plane as the geometric model. Fig. 5.11 shows the PSNRs of the rendered images with respect to the number of iterations ARC performs. The improvement from no ARC (iteration number = 0) to ARC is very significant. Another interesting phenomenon is that the PSNRs improve dramatically when ARC was first applied (iteration number = 1), but then they improve very slowly when the number of iterations increases. This was unexpected but very useful for applying ARC to dynamic scenes and adapting to viewer movements.

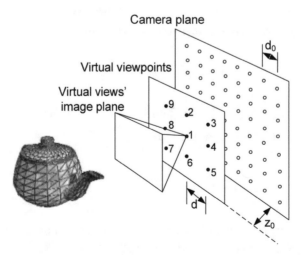

FIGURE 5.9: Setup of experiments on synthetic scenes

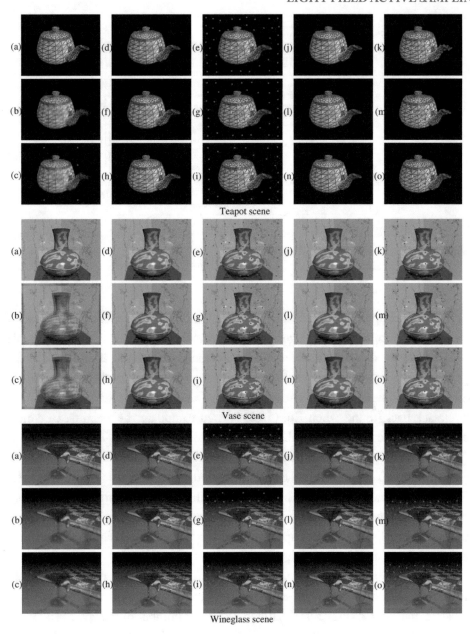

FIGURE 5.10: Results of our active rearranged capturing (ARC) algorithm. (a) Virtual view to be rendered (ground truth). (b), (c) Rendering results before ARC and projection of the camera locations on the virtual imaging plane. (d), (e) Update with Eq. (5.12), 1 iteration of ARC. (f), (g) Update with Eq. (5.12), 3 iterations of ARC. (h), (i) Update with Eq. (5.12), 10 iterations of ARC. (j), (k) Updates with Eq. (5.14), 1 iteration of ARC. (l), (m) Update with Eq. (5.14), 3 iterations of ARC. (n), (o) Update with Eq. (5.14), 10 iterations of ARC

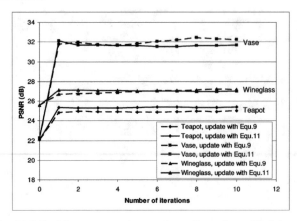

FIGURE 5.11: PSNR of ARC with respect to the number of iterations. Each iteration contains weighted VQ, camera movement and new image set capturing

Both weight updating algorithms work well and achieve similar performance. On the *teapot* scene, the second updating algorithm performs slightly better. This can be explained by Fig. 5.10, where we show some of the rendering results before and after ARC. The red dots in Figs. 5.10(c)–(o) show the projection of the camera locations to the virtual imaging plane of the rendering camera. It is interesting to observe that in *teapot*, many cameras are stuck in the pure black background when Eq. (5.12) is used to update the weights. In contrast, they immediately move to the foreground object when using Eq. (5.14). This is because although Eq. (5.12) adjusts the weights up or down exponentially, it cannot directly set the weights of those background pixels to zero like Eq. (5.14) will do. Hence the convergence of ARC with Eq. (5.12) as the updating mechanism is slower. On the other hand, after 10 iterations, the first updating algorithm achieves slightly better PSNR on *vase* and *wineglass*. This is because the second algorithm relies on a linear approximation in Eq. (5.9), which is less flexible than the first algorithm in adjusting the weights.

We next examine the performance of ARC when multiple views are rendered. In Fig. 5.12, we show PSNR curves of *teapot* and *vase* with respect to the number of virtual viewpoints and their relative distances d (Fig. 5.9). In Fig. 5.12, if the number of viewpoints is P, we use all viewpoints whose indexes are less than or equal to P in Fig. 5.9. The weight updating algorithm in Eq. (5.14) is adopted in this comparison, with three iterations of ARC. Note that in all cases capturing with ARC produces significantly higher PSNR than if no ARC is performed. On the other hand, the improvement drops when the number of viewpoints increases. This is expected because when the number of viewpoints increases, ARC needs to seek balance between multiple

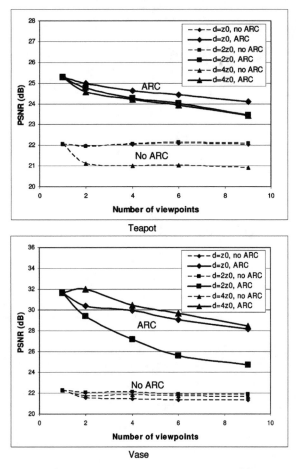

FIGURE 5.12: Performance of ARC for multiple virtual viewpoints. The symbols such as d and z_0 were defined in Fig. 5.9

views in order to achieve the best average rendering quality. Similarly, when the virtual views are more and more far apart, the to-be-rendered light rays have less overlap, which increases the difficulty for ARC to find camera locations that can render all the light rays well. Note that the above conclusions are not definite (e.g., the *vase* scene), because different viewpoints may see completely different scene objects.

CHAPTER 6

The Self-Reconfigurable Camera Array

In this chapter, we present a self-reconfigurable camera array system, which captures video sequences from an array of mobile cameras, renders novel views on the fly, and reconfigures the camera positions to achieve better rendering quality. Through this chapter, we hope to demonstrate to the reader what are the essential components of a light field/image-based rendering system, and how to assemble them together as a working system. In addition, the system serves as an example of the active rearranged sampling approach, since the cameras in our system are mounted on mobile platforms, and they can move around in order to achieve better rendering results given the virtual viewing positions.

It is worth noting that compared with a single moving camera which follows the viewer's instruction and move around, a camera array has the clear benefit that its captured images can be used by thousands of viewers to view the scene simultaneously and from arbitrary view positions. In addition, the viewers are free to move their virtual viewpoints around without worrying about the mechanical speed of the single moving camera.

6.1 SYSTEM OVERVIEW

6.1.1 Hardware

As shown in Fig. 6.1, our self-reconfigurable camera array system is composed of inexpensive off-the-shelf components. There are 48 (8 × 6) Axis 205 network cameras placed on 6 linear guides. The linear guides are 1600 mm in length; thus the average distance between cameras is about 200 mm. Vertically the cameras are 150 mm apart. They can capture up to 640 × 480 pixels images at maximally 30 fps. The cameras have built-in HTTP servers, which respond to HTTP requests and send out motion JPEG sequences. The JPEG image quality is controllable. The cameras are connected to a central computer through 100 Mbps Ethernet cables.

The cameras are mounted on a mobile platform, as shown in Fig. 6.2. Each camera is attached to a pan servo, which is a standard servo capable of rotating for about 90°. They are mounted on a platform, which is equipped with another sidestep servo. The sidestep servo is a

FIGURE 6.1: Our self-reconfigurable camera array system with 48 cameras

hacked one, and can rotate continuously. A gear wheel is attached to the sidestep servo, which allows the platform to move horizontally with respect to the linear guide. The gear rack is added to avoid slipperiness during the motion. The two servos on each camera unit allow the camera to have two degrees of freedom—pan and sidestep. However, the 12 cameras at the leftmost and rightmost columns have fixed positions and can only pan.

The servos are controlled by the Mini SSC II servo controller [38]. Each controller is in charge of no more than 8 servos (either standard servos or hacked ones). Multiple controllers can be chained; thus up to 255 servos can be controlled simultaneously through a single serial

FIGURE 6.2: The mobile camera unit

connection to a computer. In the current system, we use altogether 11 Mini SSC II controllers to control 84 servos (48 pan servos, 36 sidestep servos).

Unlike existing camera array systems such as [72, 75, 70, 76, 36], our whole system uses only one single computer (for a detailed review of existing camera arrays, please refer to [80]). The computer is an Intel Xeon 2.4 GHz dual processor machine with 1 GB of memory and a 32 MB NVIDIA Quadro2 EX graphics card. As will be detailed in Section 6.3, our rendering algorithm is so efficient that the ROI identification, JPEG image decompression, and camera lens distortion correction, which were usually performed with dedicated computers in previous systems, can all be conducted during the rendering process for a camera array at our scale. On the other hand, it is not difficult to modify our system and attribute ROI identification and image decoding to dedicated computers, as was done in the MIT distributed light field camera [75].

Fig. 6.3(a) shows a set of images about a static scene captured by our camera array. The images are acquired at 320×240 pixels. The JPEG compression quality is set to be 30 (0 being the best quality and 100 being the worst quality). Each compressed image is about 12–18 kbytes. In a 100 Mbps Ethernet connection, 48 cameras can send such JPEG image sequences to the computer simultaneously at 15–20 fps, which is satisfactory. Several problems can be spotted from these images. First, the cameras have severe lens distortions, which have to be corrected during the rendering. Second, the colors of the captured images have large variations. The Axis 205 camera does not have flexible lighting control settings. We use the "fixed indoor" white balance and "automatic" exposure control in our system. Third, the disparity between cameras is large. As will be shown later, using constant depth assumption to render the scene will generate images with severe ghosting artifacts. Finally, the captured images are noisy (Figs. 6.3(b)–(e)). Such noises are from both the CCD sensors of the cameras and the JPEG image compression. These noises bring extra challenges to the scene geometry reconstruction.

The Axis 205 cameras cannot be easily synchronized. We ensure that the rendering process will always use the most recently arrived images at the computer for synthesis. Currently, we ignore the synchronization problem during the geometry reconstruction and rendering, though it does cause problems while rendering fast moving objects, as might have been observed in the submitted companion video files.

6.1.2   Software Architecture

The system software runs as two processes, one for capturing and the other for rendering. The capturing process is responsible for sending requests to and receiving data from the cameras. The received images (in JPEG compressed format) are directly copied to some shared memory that both processes can access. The capturing process is often lightly loaded, consuming about

FIGURE 6.3: Images captured by our camera array. (a) All the images; (b)–(e) sample images from selected cameras

20% of one of the processors in the computer. When the cameras start to move, their external calibration parameters need to be calculated in real time. Camera calibration is also performed by the capturing process. As will be described in the next section, calibration of the external parameters generally runs fast (150–180 fps).

The rendering process runs on the other processor. It is responsible for ROI identification, JPEG decoding, lens distortion correction, scene geometry reconstruction, and novel view synthesis. Details about the rendering process will be described in Section 6.3.

6.2   CAMERA CALIBRATION

Since our cameras are designed to be self-reconfigurable, calibration must be performed in real time. Fortunately, the internal parameters of the cameras do not change during their motion, and can be calibrated offline. We use a large planar calibration pattern for the calibration process (Fig. 6.3). Bouguet's calibration toolbox [3] is used to obtain the internal camera parameters.

To calibrate the external parameters, we first need to extract the features on the checkerboard. We assume that the top two rows of feature points will never be blocked by the foreground objects. The checkerboard boundary is located by searching for the red strips of the board in the top region of the image. Once the left and right boundaries are identified (as shown in Fig. 6.4), we locate the top two rows of features by using a simple 5×5 linear filter:

$$
h_1 = \begin{pmatrix}
1 & 0 & 0 & 0 & -1 \\
0 & 1 & 0 & -1 & 0 \\
0 & 0 & 0 & 0 & 0 \\
0 & -1 & 0 & 1 & 0 \\
-1 & 0 & 0 & 0 & 1
\end{pmatrix}. \tag{6.1}
$$

If the output of the above filter has an absolute value larger than a threshold, it is considered as a candidate feature location. After nonmaximum suppression of the candidates, we further use

FIGURE 6.4: Locate the feature corners of the calibration pattern

a 9×9 linear filter to confirm their validity:

$$h_2 = \begin{pmatrix} 1 & 1 & 1 & 0 & 0 & 0 & -1 & -1 & -1 \\ 1 & 1 & 1 & 0 & 0 & 0 & -1 & -1 & -1 \\ 1 & 1 & 1 & 0 & 0 & 0 & -1 & -1 & -1 \\ 0 & 0 & 0 & 0 & 0 & 0 & 0 & 0 & 0 \\ 0 & 0 & 0 & 0 & 0 & 0 & 0 & 0 & 0 \\ 0 & 0 & 0 & 0 & 0 & 0 & 0 & 0 & 0 \\ -1 & -1 & -1 & 0 & 0 & 0 & 1 & 1 & 1 \\ -1 & -1 & -1 & 0 & 0 & 0 & 1 & 1 & 1 \\ -1 & -1 & -1 & 0 & 0 & 0 & 1 & 1 & 1 \end{pmatrix}. \tag{6.2}$$

A feature is valid only if the result of filter h_2 (absolute value) is also above a threshold. The above algorithm assumes that the calibration pattern is almost frontal in all the captured views. The thresholds of the two filters are chosen empirically.

The feature positions are then refined to subpixel accuracy by finding the saddle points, as in [3]. The corners below the second row are then extracted row by row. At each row, we predict the feature locations based on its previous two rows of features. The accurate positions of the features are then found through the same approach above. The results of such feature extraction are shown in Fig. 6.4. Note that if the corner detector cannot find a feature along a column for a certain row due to various reasons such as occlusions, it will stop finding features below that row in that column.

Finally, to obtain the six external parameters (three for rotation and three for translation) of the cameras, we use the algorithm proposed by Zhang [83]. The Levenberg–Marquardt method implemented in MinPack [40] is adopted for the nonlinear optimization. The above calibration process runs very fast on our processor (150–180 fps at full speed). As long as there are not too many cameras moving around simultaneously, we can perform calibration on the fly during the camera movement. In the current implementation, we constrain that at any instance at most one camera on each row can sidestep. After a camera has sidestepped, it will pan if necessary in order to keep the calibration board in the middle of the captured image.

6.3   REAL-TIME RENDERING

6.3.1   Flow of the Rendering Algorithm

In this chapter, we propose to reconstruct the geometry of the scene as a two-dimensional (2D) multiresolution mesh (MRM) with depths on its vertices, as shown in Fig. 6.5. The 2D mesh is positioned on the imaging plane of the virtual view; thus the geometry is view dependent (similar to that in [76, 60, 35]). The MRM solution significantly reduces the amount of computation spent on depth reconstruction, making it possible to be implemented efficiently in software.

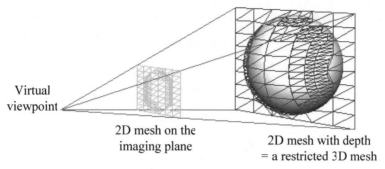

Virtual
viewpoint

2D mesh on the
imaging plane

2D mesh with depth
= a restricted 3D mesh

FIGURE 6.5: The multiresolution 2D mesh with depth information on its vertices

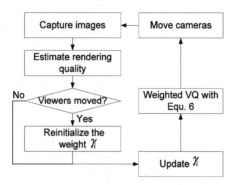

FIGURE 6.6: The flowchart of the rendering algorithm

The flowchart of the rendering algorithm is shown in Fig. 6.6. A novel view is rendered when there is an idle callback or the user moves the viewpoint. We first construct an initial sparse and regular 2D mesh on the imaging plane of the virtual view, as is shown in Fig. 6.7. For each vertex of the initial 2D mesh, we first look for a subset of images that will be used to interpolate its intensity during the rendering. Once such information has been collected, it is easy to identify the ROIs of the captured images and decode them when necessary. The depths of the vertices in the 2D mesh are then reconstructed. If a certain triangle in the mesh bears large depth variation, subdivision is performed to obtain more detailed depth information. After the depth reconstruction, the novel view can be synthesized through multitexture blending, similar to that in the unstructured Lumigraph rendering (ULR) [5]. Lens distortion is corrected in the last stage, although we also compensate the distortion during the depth reconstruction stage. Details of the proposed algorithm will be presented next.

6.3.2   Finding Close-by Images for the Mesh Vertices

Each vertex on the 2D mesh corresponds to a light ray that starts from the virtual viewpoint and passes the vertex on the imaging plane. During the rendering, it will be interpolated from

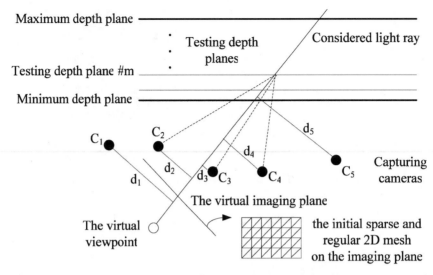

FIGURE 6.7: Locate the neighboring images for interpolation and depth reconstruction through plan sweeping

several light rays of some nearby captured images. We need to identify these nearby images for selective JPEG decoding and the scene geometry reconstruction. Unlike the ULR [5] and the MIT distributed light field camera [75] where the scene depth is known, we do not have such information at this stage, and cannot locate the neighboring images by angular differences of the light rays[1]. Instead, we adopted the distance from the cameras' center-of-projection to the considered light ray as the criterion. As shown in Fig. 6.7, the capturing cameras C_2, C_3, and C_4 have smaller distances, and will be selected as the three closest images. As our cameras are roughly arranged on a plane and point to roughly the same direction, when the scene is at a reasonably large depth, such distance is a good approximation of the angular difference used in the literature, yet it does not require the scene depth information.

6.3.3   ROI Identification and JPEG Decoding

On the initial regular coarse 2D mesh, if a triangle has a vertex that selects input image n as one of the nearby cameras, the rendering of this triangle will need image n. In other words, once all the vertices have found their nearby images, given a certain input image n, we will be able to tell what are the triangles that need it during the rendering. Such information is used to identify the ROIs of the images that need to be decoded.

---

[1] Although it is possible to find the neighboring images of the light rays for each hypothesis depth planes, we found such an approach too much time consuming.

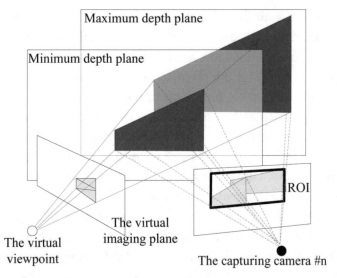

Maximum depth plane

Minimum depth plane

ROI

The virtual
imaging plane

The virtual
viewpoint

The capturing camera #n

FIGURE 6.8: Determine the ROI of a certain input image

As shown in Fig. 6.8, take image n as an example. We back-project the triangles (indicated in yellow) that need image n for rendering from the virtual imaging plane to the minimum depth plane and the maximum depth plane, and then project the resultant regions to image n. The ROI of image n is the smallest rectangular region that includes both the projected regions. Afterward, the input images that do not have an empty ROI will be JPEG decoded (partially).

6.3.4   Scene Depth Reconstruction

We reconstruct the scene depth of the light rays passing through the vertices of the 2D mesh using a plane sweeping method. Such a method has been used in a number of previous algorithms [10, 56, 75], although they all reconstruct a dense depth map of the scene. As illustrated in Fig. 6.7, we divide the world space into multiple testing depth planes. For each light ray, we assume that the scene is on a certain depth plane, and project the scene to the nearby input images obtained in Section 6.3.2. If the assumed depth is correct, we expect to see consistent colors among the projections. The plane sweeping method sweeps through all the testing depth planes, and obtain the scene depth as the one that gives the highest consistency.

Care must be taken in applying the above method. First, the locations of the depth planes should be equally spaced in the disparity space instead of in depth. Let d_{min} be the minimum scene depth, d_{max} be the maximum scene depth, and M be the number of depth planes used. The m depth plane ($m = 0, 1, \ldots, M - 1$) is located at

$$d_m = \frac{1}{\frac{1}{d_{max}} + \frac{m}{M-1}\left(\frac{1}{d_{min}} - \frac{1}{d_{max}}\right)}. \qquad (6.3)$$

Equation (6.3) is a direct result from the sampling theory by Chai et al. [8]. In the same paper they also developed a sampling theory on the relationship between the number of depth planes and the number of captured images, which is helpful in selecting the number of depth planes. Second, when projecting the test depth planes to the neighboring images, lens distortions must be corrected. Third, to improve the robustness of the color consistency matching among the noisy input images, a patch on each nearby image is taken for comparison. The patch window size relies heavily on the noise level in the input images. In our current system the input images are very noisy. We have to use an 18×18 patch window to accommodate the noise. The patch is first down-sampled horizontally and vertically by a factor of 2 to reduce some computational burden. Different patches in different input images are then compared to give an overall color consistency score. Fourth, as our cameras have large color variations, color consistency measures such as SSD do not perform very well. We applied mean-removed correlation coefficient for the color consistency verification (CCV). The normalized mean-removed inner products of all pairs of nearby input images are first obtained. Given a pair of nearby input images, e.g., i and j, the correlation coefficient of the two patches is defined as

$$r_{ij} = \frac{\sum_k (I_{ik} - \bar{I}_i)(I_{jk} - \bar{I}_j)}{\sqrt{\left[\sum_k (I_{ik} - \bar{I}_i)^2\right]\left[\sum_k (I_{jk} - \bar{I}_j)^2\right]}} \qquad (6.4)$$

where I_{ik} and I_{jk} are the kth pixel intensity in patch i and j, respectively. $\bar{I}_i$ and $\bar{I}_j$ are the mean of pixel intensities in the two patches. Equation (6.4) was widely used in traditional stereo matching algorithms [12]. The overall CCV score of the nearby input images is one minus the average correlation coefficient of all the image pairs. The depth plane resulting in the lowest CCV score will be selected as the scene depth. Since the calculation of Eq. (6.4) is very computationally expensive, we may perform early rejection if the intensities of the patches differ too much.

The depth recovery process starts with an initial regular and sparse 2D mesh, as was shown in Fig. 6.7. The depths of its vertices are obtained with the above-mentioned method. The sparse mesh with depth can serve well during the rendering if the scene does not have much depth changes. However, if the scene depth does change, a dense depth map is needed around those regions for satisfactory rendering results. We subdivide a triangle in the initial mesh if its three vertices have a large depth variation. As shown in Fig. 6.9, let the depths of a triangle's three vertices be d_{m_1}, d_{m_2}, and d_{m_3}, where m_1, m_2, m_3 are the indices of the depth planes. We subdivide this triangle if

$$\max_{p,q \in \{1,2,3\}, p \neq q} |m_p - m_q| > T \qquad (6.5)$$

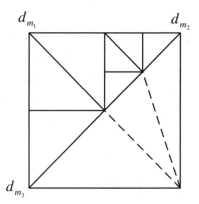

FIGURE 6.9: Subdivision of a mesh triangle

where T is a threshold set as 1 in the current implementation. During the subdivision, the midpoint of each edge of the triangle is selected as the new vertices, and the triangle is subdivided into four smaller ones. The depths of the new vertices are reconstructed under the constraints that they have to use neighboring images of the three original vertices, and their depth search range is limited to the minimum and maximum depth of the original vertices. Other than Eq. (6.5), the subdivision may also stop if the subdivision level reaches a certain preset limit.

Real-time, adaptive conversion from dense depth map or height field to a mesh representation has been studied in the literature [33, 6]. However, these algorithms assumed that a dense depth map or height field was available beforehand. Therefore, they are similar to the freeform sampling solutions discussed in Section 4.2. In contrast, our algorithm reconstructs a multiresolution mesh model directly during the rendering, and belongs to the active incremental sampling. Szeliski and Shum [62] proposed a motion estimation algorithm based on a multiresolution representation using quadtree splines, which could be considered as active incremental sampling too. Two-dimensional splines are used to interpolate between control points; while we connect the mesh vertices with triangles. They used the residual flow for the subdivision of quadtree, which is applicable for two-view stereo. In comparison, our method is for multiviews. In addition, the geometry we reconstruct is view dependent, while the quadtree splines based method is for one of the stereo views.

The size of the triangles in the initial regular 2D mesh cannot be too large, since otherwise we may miss certain depth variations in the scene. A rule of thumb is that the size of the initial triangles/grids should match that of the object features in the scene. In the current system, the initial grid size is about 1/25 of the width of the input images. Triangle subdivision is limited to no more than two levels.

6.3.5   Novel View Synthesis

After the multiresolution 2D mesh has been obtained, novel view synthesis is easy. Our rendering algorithm is very much similar to the one in the ULR [5] except that our imaging plane has already been triangulated. The basic idea of ULR is to assign weights to nearby images for each vertex, and render the scene through multitexture blending. In our implementation, the weights of the nearby images are assigned as in Eq (5.2). During the multitexture blending, only the ROIs of the input images will be used to update the texture memory when a novel view is rendered. As the input images of our system have severe lens distortions, we cannot use the 3D coordinates of the mesh vertices and the texture matrix in graphics hardware to specify the texture coordinates as in [5]. Instead, we perform the projection with lens distortion correction ourselves and provide 2D texture coordinates to the rendering pipeline. Fortunately, such projections to the nearby images have already been calculated during the depth reconstruction stage and can simply be reused.

6.3.6   Rendering Results on Synthetic Scenes

We first report some rendering results of the proposed algorithm on some synthetic static scenes, *wineglass* and *skyvase*, as shown in Fig. 6.10. Both scenes are synthesized with POV-Ray [50]. They have 64 input images, arranged regularly on a 2D plane as the light field. Wineglass exhibits huge depth variation, transparency, shadowing, and heavy occlusion; thus it is very challenging for geometry reconstruction. The skyvase scene features two semireflective walls which have their own textures and also reflect the vase in the foreground. Theoretically, no geometry reconstruction algorithm could work for this scene, because there are virtually two depths existing on the wall.

Figs. 6.10(a)–(c) are the rendering results when using a sparse mesh, a dense depth map, and an adaptive mesh, all being view dependent. The rendering position is at the center of four nearby capturing cameras (forming a rectangle) on the camera plane. As synthetic scenes have no noise, we use patch window 5×5 during the geometry reconstruction. Eight depth planes are used here for plane sweeping. Parts (d)–(f) show the geometric models we reconstructed. Both (b) and (c) have very good rendering quality, thanks to the view-dependent geometry reconstruction and the local color consistency verification. Most importantly, we reconstruct depth for 5437 vertices for the wineglass scene and 4507 vertices for the skyvase scene in the adaptive mesh approach, which are both much fewer than 76,800 vertices if a per-pixel dense depth map is reconstructed.

6.3.7   Rendering Results on Real-World Scenes

We have used our camera array system to capture a variety of scenes, both static and dynamic. The speed of the rendering process is about 4–10 fps, depending on many factors such as the

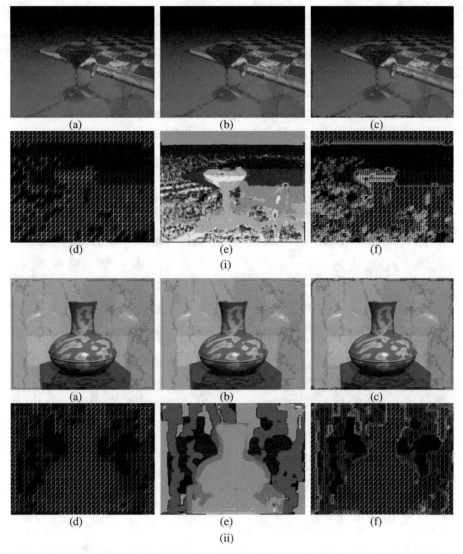

FIGURE 6.10: Synthetic scenes rendered with our proposed algorithm. (i) Scene *wineglass*; (ii) scene *skyvase*; (a),(d) use a sparse mesh to render; (b),(e) use a per-pixel depth map to render; (c),(f) use the proposed adaptive mesh to render

number of testing depth planes used for plane sweeping, the patch window size for CCV, the initial coarse regular 2D mesh grid size, the number of subdivision levels used during geometry reconstruction, and the scene content. For the scenes we have tested, the above parameters can be set to fixed values. For instance, our default setting is 12 testing depth planes for depth sweeping, 18×18 patch window size, $1/25$ of the width of the input images as initial grid size, and maximally two levels of subdivision.

The time spent on each step of the rendering process under default setting is as follows. Finding neighboring images and ROI of them takes less than 10 ms. JPEG decoding takes 15–40 ms. Geometry reconstruction takes about 80–120 ms. New view synthesis takes about 20 ms.

The rendering results of some static scenes are shown in Fig. 6.11. In these results the cameras are evenly spaced on the linear guide. The rendering positions are roughly on the camera plane but not too close to any of the capturing cameras. Figs. 6.11(a)–(c) are results rendered with the constant depth assumption. The ghosting artifacts are very severe, because the spacing

FIGURE 6.11: Scenes captured and rendered with our camera array. (i) Scene *toys*; (ii) scene *train*; (iii) scene *girl and checkerboard*; (iv) scene *girl and flowers*; (a) rendering with a constant depth at the background; (b) rendering with a constant depth at the middle object; (c) rendering with a constant depth at the closest object; (d) rendering with the proposed method; (e) multiresolution 2D mesh with depth reconstructed on the fly; brighter intensity means smaller depth

between our cameras is larger than most previous systems [75, 41]. Fig. 6.11(d) is the result from the proposed algorithm. The improvement is significant. Fig. 6.11(e) shows the reconstructed 2D mesh with depth information on its vertices. The grayscale intensity represents the depth— the brighter the intensity, the closer the vertex. Like many other geometry reconstruction algorithms, the geometry we obtained contains some errors. For example, in the background region of scene *toys*, the depth should be flat and far, but our results have many small "bumps". This is because part of the background region has no texture, which is prone to error for depth recovery. However, the rendered results are not affected by these errors because we use view-dependent geometry and the local color consistency always holds at the viewpoint.

Fig. 6.12 gives the comparison of the rendering results using a dense depth map and our adaptive mesh, similar to that in Fig. 6.10 but for real-world scenes. Again, using adaptive mesh produces rendering images of almost the same quality as using dense depth map, but at a much smaller computational cost.

6.3.8   Discussions

Our current system has certain hardware limitations. For example, the images captured by the cameras are at 320×240 pixels and the image quality is not very high. This is mainly constrained by the throughput of the Ethernet cable. Upgrading the system to gigabit Ethernet or using more computers to handle the data could solve this problem. For dynamic scenes, we notice that our system cannot catch up with very fast moving objects. This is due to the fact that the cameras are not synchronized.

We find that when the virtual viewpoint moves out of the range of the input cameras, the rendering quality degrades quickly. A similar effect was reported in [75, 63]. The poor extrapolation results are due to the lack of scene information in the input images during the geometry reconstruction.

Since our geometry reconstruction algorithm resembles the traditional window-based stereo algorithms, they share some limitations. For instance, when the scene has large depth discontinuity, our algorithm does not perform very well along the object boundary (especially when both foreground and background objects have strong textures). In the current implementation, our correlation window has very large size (18×18) in order to tolerate the noisy input images. Such a big correlation window tends to smooth the depth map. Figs. 6.13(i-d) and (iii-d) show the rendering results of two scenes with large depth discontinuity. Note the artifacts around the boundaries of the objects. To solve this problem, one may borrow ideas from the stereo literature [22, 24], which will be our future work. Alternatively, since we have built a reconfigurable camera array, we may reconfigure the arrangement of the cameras, as will be described in the next section.

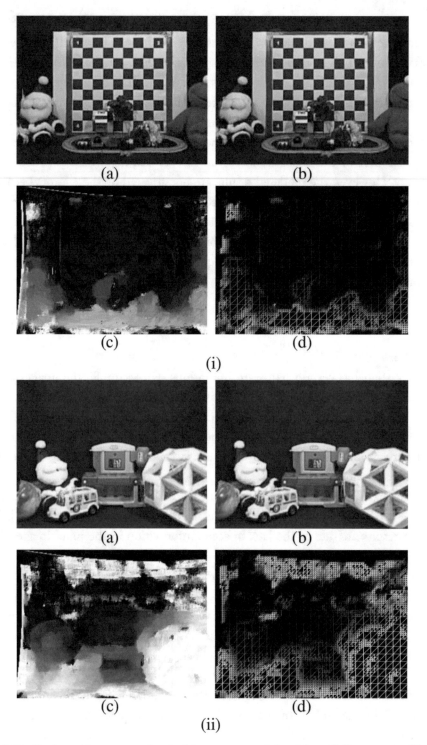

FIGURE 6.12: Real-world scenes rendered with our proposed algorithm. (i) Scene *train*; (ii) scene *toys*; (a),(c) use a per-pixel depth map to render; (b),(d) use the proposed adaptive mesh to render

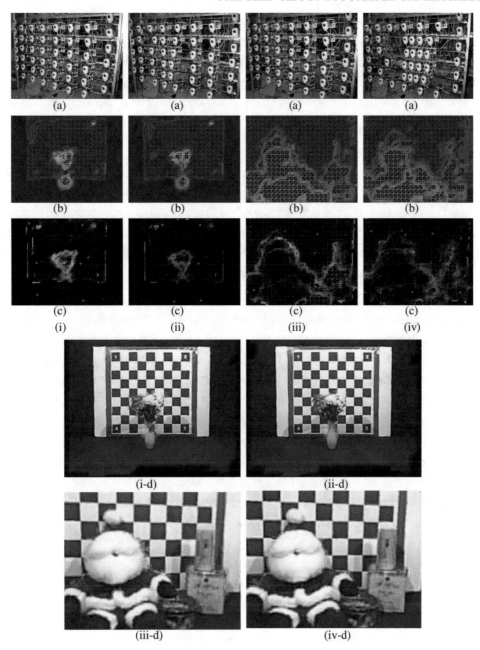

FIGURE 6.13: Scenes rendered by reconfiguring our camera array. (i) Scene *flower*; cameras are evenly spaced; (ii) scene *flower*; cameras are self-reconfigured (6 epochs); (iii) scene *Santa*; cameras are evenly spaced; (iv) scene *Santa*; cameras are self-reconfigured (20 epochs); (a) the camera arrangement; (b) reconstructed depth map—brighter intensity means smaller depth; (c) the CCV score of the mesh vertices and the projection of the camera positions to the virtual imaging plane (red dots)—darker intensity means better consistency; (d) rendered image

6.4   SELF-RECONFIGURATION OF THE CAMERAS

In Section 5.3 we have presented an active rearranged sampling algorithm for a light field. In that work we assumed that all the capturing cameras can move freely on the camera plane. Such assumption is very difficult to implement in practical systems. In this section, we present a revised algorithm for the self-reconfiguration of the cameras, given that they are constrained on the linear guides.

6.4.1   The Proposed Local Rearrangement Algorithm

Figs. 6.13(i-c) and (iii-c) show the local inconsistency score obtained while reconstructing the scene depth (Section 6.3.4). They are directly used for our active rearranged sampling algorithm. It is obvious that if the consistency is bad (high score), the reconstructed depth tends to be wrong, and the rendered scene tends to have low quality. Our camera self-reconfiguration (CSR) algorithm will thus move the cameras to where the score is high.

Our CSR algorithm contains the following steps:

1. *Locate the camera plane and the linear guides* (as line segments on the camera plane). The camera positions in the world coordinates are obtained through the calibration process. Although they are not strictly on the same plane, we use an approximated one which is parallel to the checkerboard. The linear guides are located by averaging the vertical positions of each row of cameras on the camera plane. As shown in Fig. 6.14, denote the vertical coordinates of the linear guides on the camera plane as Y_j, $j = 1, \ldots, 6$.

2. *Back-project the vertices of the mesh model to the camera plane.* Although during the depth reconstruction the mesh can be subdivided, during this process we only make use of the initial sparse mesh (Fig. 6.7). In Fig. 6.14, one mesh vertex was back-projected as (x_i, y_i) on the camera plane. Note that such back-projection can be performed even if there are multiple virtual views to be rendered, thus the proposed CSR algorithm is applicable to situations where there exist multiple virtual viewpoints.

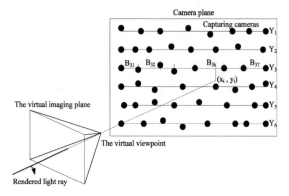

FIGURE 6.14: Self-reconfiguration of the cameras

3. *Collect CCV score for each pair of neighboring cameras on the linear guides.* The capturing cameras on each linear guide naturally divide the guide into seven segments. Let them be B_{jk}, where j is the row index of the linear guide, k is the index of bins on that guide, $1 \leq j \leq 6$, $1 \leq k \leq 7$. If a back-projected vertex (x_i, y_i) satisfies

$$Y_{j-1} < y_i < Y_{j+1} \quad \text{and} \quad x_i \in B_{jk}, \qquad (6.6)$$

the CCV score of the vertex is added to bin B_{jk}. After all the vertices have been back-projected, we obtain a set of accumulated CCV scores for each linear guide, denoted as S_{jk}, where j is the row index of the linear guide, and k is the index of bins on that guide.

4. *Determine which camera to move on each linear guide.* Given a linear guide j, we look for the largest S_{jk}, $1 \leq k \leq 7$. Let it be S_{jK}. If the two cameras forming the corresponding bin B_{jK} are not too close to each other, one of them will be moved toward the other (thus reducing their distance). Note that each camera is associated with two bins. To determine which one of the two cameras should move, we check their other associated bin and move the camera with a smaller accumulated CCV score in its other associated bin.

5. *Move the cameras.* Once the moving cameras have been decided, we issue them commands such as "move left" or "move right"[2]. Once the cameras are moved, the process waits until it is confirmed that the movement has finished and the cameras are recalibrated. Then it jumps back to step 1 for the next epoch of movement.

6.4.2   Results

We show results of the proposed CSR algorithm in Fig. 6.13. In Figs. 6.13(i) and (iii), the capturing cameras are evenly spaced on the linear guide. Fig. 6.13(i) is rendered behind the camera plane, and Fig. 6.13(iii) is rendered in front of the camera plane. Due to depth discontinuities, some artifacts can be observed from the rendered images (Figs. 6.13(i-d) and (iii-d)) along the object boundaries. Fig. 6.13(b) is the reconstructed depth of the scene at the virtual viewpoint. Fig. 6.13(c) is the CCV score obtained during the depth reconstruction. It is obvious that along the object boundaries, the CCV score is high, which usually means wrong/uncertain reconstructed depth, or bad rendering quality. The red dots in Fig. 6.13(c) are the projections of the capturing camera positions to the virtual imaging plane.

---

[2]We can only send such commands to the sidestep servos, because the servos were hacked for continuous rotation. The positions of the cameras after movement are unpredictable, and can only be obtained through the calibration process.

Figs. 6.13 (ii) and (iv) show the rendering result after CSR. Fig. 6.13(ii) is the result of 6 epochs of camera movement, and Fig. 6.13(iv) is after 20 epochs. It can be seen from the CCV score map (Fig. 6.13(c)) that after the camera movement, the consistency generally gets better. The cameras have been moved, which is reflected as the red dots in Fig. 6.13(c). The cameras move toward the regions where the CCV score is high, which effectively increases the sampling rate for the rendering of those regions. Figs. 6.13(ii-d) and (iv-d) show the rendering results after self-reconfiguration, which is much better than Figs. 6.13(i-d) and (iii-d).

6.4.3   Discussions

The major limitation of the self-reconfigurable camera array is that the motion of the cameras is generally slow. When the computer writes a command to the serial port, the command will be buffered in the Mini SSC II controller for ~15 ms before being sent to the servo. After the servo receives the command, there is also a long delay (hundreds of milliseconds) before it moves enough distance. Therefore, during the self-reconfiguration of the cameras, we have to assume that the scene is either static or moving very slowly, and the viewer is not changing his/her viewpoint all the time. During the motion of the cameras, since the calibration process and the rendering process run separately, we observe some jittering artifacts of the rendered images when the moved cameras have not been fully calibrated.

There is no collision detection in the current system while moving the cameras. Although the calibration process is very stable and gives fairly good estimation of the camera positions, collision could still happen. In Section 6.4.1, we have a threshold for verifying whether two cameras are too close to each other. The current threshold is set as 10 cm, which is reasonably safe in all our experiments.

CHAPTER 7

Conclusions and Future Work

This book sets out to address the following two questions:

> *How many images are needed for IBR? If such a number is limited, where shall we capture these images?*

In this final chapter let us summarize the work we have presented in this book.

Uniform IBR sampling analysis. We presented a new method to analyze the Fourier spectrum of IBR scenes, which is able to handle both non-Lambertian surface and occlusions. We showed that in both cases the required sampling rate is higher than Lambertian and nonoccluded ones. Considering that the IBR sampling problem is a multidimensional sampling problem, we also applied the generalized sampling theorem for IBR sampling. We are able to reduce the sampling rate by a factor of 50% in theory, and achieve better rendering quality for complex scenes. We also concluded that rectangular sampling is still preferable for most scenes thanks to its simplicity.

A very general framework on freeform sampling and active sampling. Compared with the traditional uniform sampling theorem, the freeform sampling framework has more practical considerations such as the reconstruction method, the reconstruction set, the sampling noise, etc. General solutions of freeform sampling were described in this dissertation, including decremental sampling, incremental sampling, and rearranged sampling. We also presented active sampling as a special case of freeform sampling, where the function values of the sampled signal on the reconstruction set are unknown. We applied it to the light field and designed several algorithms, which demonstrated that active sampling is superior to the traditional uniform sampling method.

The self-reconfigurable camera array. We presented a self-reconfigurable camera array, where the cameras are mobile. We developed a very efficient algorithm for the real-time rendering of dynamic scenes. Active sampling was widely used in the algorithm to improve the rendering speed. The source code of the rendering algorithm was distributed online[1] to inspire more work

---

[1] http://amp.ece.cmu.edu/projects/MobileCamArray/

along this direction. We also showed that by moving the cameras around for active sampling, we can improve the rendering quality, especially at object boundaries.

This book opens up some new interesting directions for further research on various topics, which are as follows:

Light field sampling analysis with scene geometry. When part of the scene geometry is known or reconstructed, it is not clear what the real minimum sampling rate is. Theoretically, the minimum sampling rate should be determined only by the scene content. However, as discussed in Chapter 4, in practice, the sampling rate will be determined by the reconstruction/rendering algorithm as well. On the other hand, if we are free to apply the best possible reconstruction/ rendering algorithm, what will be the minimum sampling rate? Discovery of such theory is very important for guiding the construction of practical IBR systems.

Real-time processing of dynamic IBR data. While the extension of IBR from static scenes to dynamic scenes seems straightforward, many new research problems arise. For instance, dynamic IBR requires all its stages to be "real time", as the scene is constantly changing. This includes real-time capturing, real-time storage, real-time calibration, real-time tracking, real-time rendering, real-time compression/streaming, etc. Some of these problems have already been studied in the literature. For example, the Stanford multicamera array [72] is capable of capturing and storing videos from many cameras at 30 fps to SCSI disk arrays. They used a dedicated hardware board to perform MPEG-2 compression of the captured videos. A relatively complete survey on IBR compression algorithms is available in [80]. Online streaming of light field/concentric mosaics has been studied in [81, 82, 52]. In Chapter 6, we have discussed some solutions for the real-time calibration and rendering from multiple mobile cameras. However, more work is needed, such as a better geometry reconstruction algorithm for heavily occluded scenes. Using our camera array to track moving objects is another problem that is very interesting.

Vision sensor network. We strongly believe that it is beneficial to add more and more function- alities to the sensor (camera), such as compression, networking, and mobility. Such migration in functionality from the central computer to the sensors not only reduces the load of the central computer, making the whole system more scalable, but also allow the sensors to distribute in a wider area, making it a true vision sensor network. In practice, due to bandwidth constraints, we expect to have limited resolution on the vision sensors. Synthesizing high-quality novel views from such low-resolution vision sensors is a new problem, and may borrow ideas from the superresolution literature. The active sampling framework proposed in this dissertation may also be used to figure out the best distribution of these vision sensors.

Multiview image/video processing. The multiple views captured for a scene can not only be used to perform rendering but also many other tasks. Most current image/video processing research topics can benefit from the availability of multiple views of the same object. To name a few, they include image/video compression, image/video restoration, image/video segmentation, image/video scene analysis, pattern recognition, etc.

Other applications of active sampling.   As a very general framework, active sampling may be used in many other applications. Recently there has been increasing interest in the application of active sampling in information retrieval. We have proposed to use active learning for hidden annotation [78]. Tong and Chang have also applied support vector machine (SVM) based active learning for obtaining the user's concept of query [66]. Naphade et al. [43] also used SVM for the active annotation of video databases. Another application of active sampling may be in the area of computer tomography (CT). In CT, one circle of scanning might only cover a small area of the scene in order to improve the precision. Due to the huge cost of storage and reconstruction, active sampling can be applied to select regions that are of most importance, or guarantee that all the scanned regions have the same reconstruction quality. In image-based relighting [46], active sampling may be a good choice to reduce the number of light patterns applied for the scene in order to achieve relighting from arbitrary lighting conditions.

References

[1]   E. H. Adelson and J. R. Bergen, "The plenoptic function and the elements of early vision," in Computational Models of Visual Processing, M. Landy and J. A. Movshon, (Eds.), Cambridge, MA: MIT Press, 1991, pp. 3–20.

[2]   A. Aldroubi and K. Gröchenig, "Nonuniform sampling and reconstruction in shift-invariant spaces," *SIAM Rev.*, Vol. 43, No. 4, pp. 585–620, 2001. doi:10.1137/S0036144501386986

[3]   J.-Y. Bouguet, *Camera Calibration Toolbox for Matlab*, http://www.vision.caltech.edu/bouguetj/calib_doc/, 1999.

[4]   A. Broadhurst and R. Cipolla, "A statistical consistency check for the space carving algorithm," in *Proc. 11th British Machine Vision Conference*, 2000.

[5]   C. Buehler, M. Bosse, L. McMillan, S. J. Gortler, and M. F. Cohen, "Unstructured lumigraph rendering," in *Proc. SIGGRAPH*, 2001, pp. 425–432.

[6]   S. Sethuraman, C. B. Bing, and H. S. Sawhney, "A depth map representation for real-time transmission and view-based rendering of a dynamic 3D scene," in *1st Int. Symp. on 3D Data Processing Visualization and Transmission*, 2002.

[7]   E. Camahort, A. Lerios, and D. Fussell, "Uniformly sampled light fields," in *9th Eurographics Workshop on Rendering*, 1998.

[8]   J.-X. Chai, S.-C. Chan, H.-Y. Shum, and X. Tong, "Plenoptic sampling," in *Proc. SIGGRAPH*, 2000, pp. 307–318.

[9]   D. A. Cohn, Z. Ghahramani, and M. I. Jordan, "Active learning with statistical models," *J. Artif. Intell. Res.* Vol. 4, pp. 129–145, 1996.

[10] R. T. Collins, "A space-sweep approach to true multi-image matching," in *Proc. CVPR*, 1996.

[11] D. E. Dudgeon and R. M. Mersereau, *Multidimensional Digital Signal Processing*, (Signal Processing Series). Englewood Cliffs, NJ: Prentice-Hall, 1984.

[12] O. Faugeras, B. Hotz, H. Mathieu, T. Viéville, Z. Zhang, P. Fua, E. Théron, L. Moll, G. Berry, J. Vuillemin, P. Bertin, and C. Proy, "Real time correlation-based stereo: Algorithm, implementations and applications," Technical Report 2013, INRIA, 1993.

[13] S. Fleishman, D. Cohen-Or, and D. Lischinski, "Automatic camera placement for image-based modeling," in *Computer Graphics Forum*, 1999.

[14] P. Fränti, J. Kivijärvi, and O. Nevalainen, "Tabu search algorithm for codebook generation in vector quantization," *Pattern Recognit.*, Vol. 31, No. 8, pp. 1139–1148, 1998. doi:10.1016/S0031-3203(97)00127-1

[15] M. R. Garey, D. S. Johnson, and H. S. Witsenhausen, "The complexity of the generalized Lloyd–Max problem," *IEEE Trans. Inform. Theory*, Vol. 28, No. 2, pp. 255–256, 1982. doi:10.1109/TIT.1982.1056488

[16] A. Gershun, "The light field," *J. Math. Phys.*, Vol. 18, pp. 51–151, 1939, Translated by P. Moon and G. Timoshenko.

[17] S. J. Gortler, R. Grzeszczuk, R. Szeliski, and M. F. Cohen, "The lumigraph," in *Proc. SIGGRAPH*, 1996, pp. 43–54.

[18] M. Hasenjäger, H. Ritter, and K. Obermayer, "Active learning in self-organizing maps," in *Kohonen Maps*, E. Oja and S. Kaski, Eds., Amsterdam: Elsevier, 1999, pp. 57–70.

[19] H. Hoppe, "Progressive meshes," *Proc. SIGGRAPH*, 1996, pp. 99–108.

[20] I. Ihm, S. Park, and R. Lee, "Rendering of spherical light fields," in *Pacific Graphics*, 1997.

[21] CBS Broadcasting Inc., http://www.cbs.com/.

[22] T. Kanade and M. Okutomi, "A stereo matching algorithm with an adaptive window: theory and experiment," *IEEE Trans. Pattern Anal. Mach. Intell.*, Vol. 16, No. 9, pp. 920–932, 1994. doi:10.1109/34.310690

[23] S. B. Kang, "A survey of image-based rendering techniques," Technical Report CRL 97/4, Cambridge Research Lab, Cambridge, MA, 1997.

[24] S. B. Kang, R. Szeliski, and J. Chai, "Handling occlusions in dense multi-view stereo," in *Proc. CVPR*, 2001.

[25] T. Koga, K. Iinuma, A. Hirano, Y. Iijima, and T. Ishiguro, "Motion-compensated interframe coding for video conferencing," in *Proc. NTC*, 1981.

[26] A. Krogh and J. Vedelsby, "Neural network ensembles, cross validation, and active learning," in *Advances in Neural Information Processing Systems*, Vol. 7, G. Tesauro, D. Touretzky and T. Leen, Eds., Cambridge, MA: MIT Press, 1995, pp. 231–238.

[27] M. P. Lai and W. C. Ma, "A novel four-step search algorithm for fast block motion estimation," *IEEE Trans. Circuits Syst. Video Technol.*, Vol. 6, No. 3, pp. 313–317, 1996. doi:org/10.1109/76.499840

[28] H. Lensch, "Techniques for hardware-accelerated light field rendering," Master's thesis, Friedrich-Alexander-Universität Erlangen-Nürnberg, 1999.

[29] M. Levoy and P. Hanrahan, "Light field rendering," in *Proc. SIGGRAPH*, 1996, pp. 31–42.

[30] Z. C. Lin and H. Y. Shum, "On the number of samples needed in light field rendering with constant-depth assumption," in *Proc. CVPR*, 2000. doi:10.1023/A:1020153824351

[31] Z.-C. Lin, T.-T. Wong, and H.-Y. Shum, "Relighting with the reflected irradiance field: representation, sampling and reconstruction," Int. *J. Comput. Vis.*, Vol. 49, No. 2–3, pp. 229–246, 2002.

[32] Y. Linde, A. Buzo, and R. Gray, "An algorithm for vector quantizer design," *IEEE Trans. Commun.*, Vol. 28, No. 1, pp. 84–95, 1980. doi:10.1109/TCOM.1980.1094577

[33] P. Lindstrom, D. Koller, W. Ribarsky, L. F. Hodges, and N. Faust, "Real-time, continuous level of detail rendering of height fields," in *Proc. SIGGRAPH*, 1996, pp. 109–118.

[34] D. Marchand-Maillet, "Sampling theory for image-based rendering," Master's thesis, EPFL, 2001.

[35] W. Matusik, C. Buehler, and L. McMillan, "Polyhedral visual hulls for real-time rendering," in *Proc. Eurographics Workshop on Rendering*, 2001.

[36] W. Matusik, C. Buehler, R. Raskar, S. Gortler, and L. McMillan, "Image-based visual hulls," in *Proc. SIGGRAPH*, 2000, pp. 369–374.

[37] G. Miller, S. Rubin, and D. Ponceleon, "Lazy decompression of surface light fields for precomputed global illumination," in *Eurographics Rendering Workshop*, 1998.

[38] MiniSSC-II, Scott Edwards Electronics Inc., http://www.seetron.com/ssc.htm.

[39] J. L. Mitchell, W. B. Pennebaker, C. E. Fogg, and D. J. LeGall, *Mpeg video: Compression Standard*. Dordrecht: Kluwer, 1996.

[40] J. J. Moré, "The Levenberg–Marquardt algorithm, implementation and theory," in *Numerical Analysis*, (Lecture Notes in Mathematics 630), G. A. Watson, Eds., Barlin: Springer, 1977, pp. 105–116.

[41] T. Naemura, J. Tago, and H. Harashima, "Real-time video-based modeling and rendering of 3d scenes," *IEEE Comput. Graphics Appl.*, Vol. 22, No. 2, pp. 66–73, 2002. doi:10.1109/38.988748

[42] R. Namboori, H. C. Tech, and Z. Huang, "An adaptive sampling method for layered depth image," in *Computer Graphics International (CGI)*, 2004.

[43] M. R. Naphade, C. Y. Lin, J. R. Smith, B. Tseng, and S. Basu, "Learning to annotate video databases," in *SPIE Conf. on Storage and Retrieval on Media databases*, 2002.

[44] M. Oliveira, "Image-based modeling and rendering techniques: a survey," *Rev. Inform. Teór. Apl.*, Vol. 9, No. 2, pp. 37–66, 2002.

[45] J. O'Rourke, "Art gallery theorems and algorithms," in *The International Series of Monographs on Computer Science*. Oxford: Oxford University Press, 1987.

[46] P. Peers and P. Dutré, "Wavelet environment matting," in *Proc. 14th Eurographics Workshop on Rendering*, 2003.

[47] W. B. Pennebaker and J. L. Mitchell, "*JPEG: Still Image Data Compression Standard*, 1st edition. Dordrecht: Kluwer, 1993.

[48] B. T. Phong, "Illumination for computer generated pictures," *Commun. ACM*, Vol. 18, No. 6, pp. 311–317, 1975. doi:org/10.1145/360825.360839

[49] R. Pito, "A solution to the next best view problem for automated surface acquisition," *IEEE Trans. Pattern Anal. Mach. Intell.*, Vol. 21, No. 10, pp. 1016–1030, 1999. doi:org/10.1109/34.799908

[50] Pov-Ray, http://www.povray.org.

[51] R. Ramamoorthi and P. Hanrahan, "A signal-processing framework for inverse rendering," in *Proc. SIGGRAPH*, 2001, pp. 117–128.

[52] P. Ramanathan, M. Kalman, and B. Girod, "Rate-distortion optimized streaming of compressed light fields," in *Proc. ICIP*, 2003.

[53] M. K. Reed, "Solid model acquisition from range images," Ph.D. thesis, Columbia University, 1998.

[54] A. Said and W. A. Pearlman, "A new fast and efficient image codec based on set partitioning in hierarchical trees," *IEEE Trans. Circuits Syst. Video Technol.*, Vol. 6, No. 3, pp. 243–250, 1996.

[55] H. Schirmacher, W. Heidrich, and H. P. Seidel, "Adaptive acquisition of lumigraphs from synthetic scenes," in *Proc. EUROGRAPHICS*, 1999.

[56] S. M. Seitz and C. R. Dyer, "Photorealistic scene reconstruction by voxel coloring," in *Proc. CVPR*, 1997.

[57] J. Shade, S. Gortler, L.-W. He, and R. Szeliski, "Layered depth images," in *Proc. SIGGRAPH*, 1998, pp. 231–242.

[58] H.-Y. Shum and L.-W. He, "Rendering with concentric mosaics," *Proc. SIGGRAPH*, 1999, pp. 299–306.

[59] H.-Y. Shum, S.B. Kang, and S.-C. Chan, "Survey of image-based representations and compression techniques," *IEEE Trans. Circuits Syst. Video Technol.*, Vol. 13, No. 11, pp. 1020–1037, 2003. doi:org/10.1109/TCSVT.2003.817360

[60] G. G. Slabaugh, R. W. Schafer, and M. C. Hans, "Image-based photo hulls," Technical Report HPL-2002-28, HP Labs, 2002.

[61] P. P. Sloan, M. F. Cohen, and S. J. Gortler, "Time critical lumigraph rendering," in *Symp. on Interactive 3D Graphics*, 1997.

[62] R. Szeliski and H.-Y. Shum, "Motion estimation with quadtree splines," *IEEE Trans. Pattern Anal. Mach. Intell.*, Vol. 18, No. 12, pp. 1199–1210, 1996. doi:org/10.1109/34.546257

[63] R. Szeliski, "Prediction error as a quality metric for motion and stereo," in *Proc. ICCV*, 1999.

[64] D. S. Taubman and M. W. Marcellin, *JPEG2000: Image Compression Fundamentals, Standards, and Practice*. Dordrecht: Kluwer, 2001.

[65] Carnegie Mellon Goes to the Super Bowl, http://www.ri.cmu.edu/events/sb35/tksuperbowl.html.

[66] S. Tong and E. Chang, "Support vector machine active learning for image retrieval," in *Proc. ACM Multimedia*, 2001.

[67] M. Unser, "Sampling—50 years after shannon," *Proc. IEEE*, Vol. 88, No. 4, pp. 569–587, 2000. doi:10.1109/5.843002

[68] P. P. Vaidyanathan and T. Q. Nguyen, "Eigenfilters: a new approach to least-squares fir filter design and applications including nyquist filters," *IEEE Trans. CAS*, Vol. 34, No. 1, pp. 11–23, 1987.

[69] J. Vaisey and A. Gersho, "Simulated annealing and codebook design," in *Proc. ICASSP*, 1988.

[70] S. Vedula, S. Baker, and T. Kanade, "Spatio-temporal view interpolation," in *13th ACM Eurographics Workshop on Rendering*, 2002.

[71] T. Werner, V. Hlaváč, A. Leonardis, and T. Pajdla, "Selection of reference views for image-based representation," in *Proc. ICPR*, 1996.

[72] B. Wilburn, M. Smulski, H.-H. K. Lee, and M. Horowitz, "The light field video camera," in *Proc. Media Processors*, 2002.

[73] T.-T. Wong, C. W. Fu, P.-A. Heng, and C.-S. Leung, "The plenoptic illumination function," *IEEE Trans. Multimedia*, Vol. 4, No. 3, 361–371. doi:org/10.1109/TMM.2002.802835

[74] D. N. Wood, D. I. Azuma, K. Aldinger, B. Curless, T. Duchamp, D. H. Salesin, and W. Stuetzle, "Surface light fields for 3d photography," in *Proc. SIGGRAPH*, 2000, pp. 287–296.

[75] J. C. Yang, M. Everett, C. Buehler, and L. McMillan, "A real-time distributed light field camera," *Eurographics Workshop on Rendering*, 2002.

[76] R. Yang, G. Welch, and G. Bishop, "Real-time consensus-based scene reconstruction using commodity graphics hardware," *Proc. Pacific Graphics*, 2002.

[77] C. Zhang, "On sampling of image-based rendering data," Ph.D. thesis, Department of Electrical and Computer Engineering, Carnegie Mellon University, 2004. doi:org/10.1109/TMM.2002.1017738

[78] C. Zhang and T. Chen, "An active learning framework for content-based information retrieval," *IEEE Trans. Multimedia*, Vol. 4, No. 2, pp. 260–268, 2002.

[79] ——, "A system for active image-based rendering," in *Proc. ICME*, 2003.

[80] ——, "A survey on image-based rendering—representation, sampling and compression," *Signal Process. Image Commun.*, Vol. 19, No. 1, pp. 1–28, 2004. doi:org/10.1016/j.image.2003.07.001

[81] C. Zhang and J. Li, "Compression of lumigraph with multiple reference frame (MRF) prediction and just-in-time rendering," in *IEEE Data Compression Conference*, 2000.

[82] ——, "Interactive browsing of 3D environment over the internet," *Proc. VCIP*, 2001.

[83] Z. Zhang, "A flexible new technique for camera calibration," Technical Report, MSR-TR-98-71, 1998.

Printed in the United States
by Baker & Taylor Publisher Services